The Technology of Property Rights

The Political Economy Forum

Sponsored by PERC
Series editor: Terry L. Anderson

The Political Economy of the American West
edited by Terry L. Anderson and Peter J. Hill
Property Rights and Indian Economies
edited by Terry L. Anderson
Public Lands and Private Rights: The Failure of Scientific Management
by Robert H. Nelson
Taking the Environment Seriously
edited by Roger E. Meiners and Bruce Yandle
Wildlife in the Marketplace
edited by Terry L. Anderson and Peter J. Hill
The Privatization Process: A Worldwide Perspective
edited by Terry L. Anderson and Peter J. Hill
Enviro-Capitalists: Doing Good While Doing Well
by Terry L. Anderson and Donald R. Leal
Environmental Federalism
edited by Terry L. Anderson and Peter J. Hill
Defending Illusions: Federal Protection of Ecosystems
by Allan K. Fitzsimmons
The Market Meets the Environment: Economic Analysis of Environmental Policy
edited by Bruce Yandle
The Common Law and the Environment: Rethinking Statutory Basis for Modern Environmental Law
edited by Roger E. Meiners and Andrew P. Morriss
Who Owns the Environment?
edited by Roger E. Meiners and Peter J. Hill
Water Marketing: The Next Generation
edited by Terry L. Anderson and Peter J. Hill
Common Sense and Common Law for the Environment
by Bruce Yandle
A Burning Issue: A Case for Abolishing the U.S. Forest Service
by Robert H. Nelson
Conservation Reconsidered: Nature, Virtue, and American Liberal Democracy
edited by Charles T. Rubin
The Politics and Economics of Park Management
edited by Terry L. Anderson and Alexander N. James

The Technology of Property Rights

Edited by
Terry L. Anderson and Peter J. Hill

ROWMAN & LITTLEFIELD PUBLISHERS, INC.
Lanham • Boulder • New York • Oxford

ROWMAN & LITTLEFIELD PUBLISHERS, INC.

Published in the United States of America
by Rowman & Littlefield Publishers, Inc.
4720 Boston Way, Lanham, Maryland 20706
www.rowmanlittlefield.com

12 Hid's Copse Road
Cumnor Hill, Oxford OX2 9JJ, England

British Library Cataloguing in Publication Information Available

Library of Congress Cataloging-in-Publication Data

The technology of property rights / edited by Terry L. Anderson and Peter J. Hill.
p. cm. — (The political economy forum)
Includes bibliographical references and index.
ISBN 0-7425-2060-9 (cloth : alk. paper) — ISBN 0-7425-2061-7 (pbk. : alk. paper)
1. Right of property—Economic aspects. 2. Technological innovations—Economic aspects. I. Anderson, Terry Lee, 1946– II. Hill, Peter Jensen. III. Series.

HB701 .T365 2001
323.4'6—dc21

2001031988

Printed in the United States of America

∞™ The paper used in this publication meets the minimum requirements of
American National Standard for Information Sciences—Permanence of Paper
for Printed Library Materials, ANSI/NISO Z39.48-1992.

Contents

Tables and Figures

TABLES

FIGURES

Acknowledgments

The Political Economy Forum is always a bigger undertaking than the final volume reflects. First, it takes a commitment from idea entrepreneurs willing to invest in the program. In this case PERC is grateful to the Alex C. Walker Educational and Charitable Trust, the Thirty Five Twenty Foundation, the Lynde and Harry Bradley Foundation, and Dunn's Foundation for the Advancement of Right Thinking. Without their investment, the forum and book would never have happened.

Though the funding helps compensate the authors and forum participants, it is never sufficient to cover their full opportunity costs. Therefore, we are appreciative of the authors' willingness to donate some of their time and energy to these papers and their patience with our editing; we hope the reader reaps a return from it. Forum participants take their compensation in the form of intellectual stimulation, but given the quality of the discussion we know that they probably gave more than they received.

Also, we can never thank the PERC staff enough. Colleen Lane and Monica Guenther always make sure the forum comes off without a hitch, and Michelle Johnson and Sheila Spain see to the details of communication and word processing. And then there is Dianna Rienhart, whose untiring efforts in preparing the manuscript for publishing ultimately make the book letter perfect. Dianna's name should be listed on the cover as the third editor.

Introduction

The Technology of Property Rights

Terry L. Anderson and Peter J. Hill

In the past two decades one of the most significant developments in economic theory has been the integration of institutions into the neoclassical model. By institutions we mean the formal and informal rules that govern interactions among individuals. This integration has made institutions endogenous rather than exogenous. That is to say, economists, political economists, and lawyers have made progress at explaining how and when property rights evolve.

Since Harold Demsetz's (1967) article, there have been numerous theoretical and empirical contributions to our understanding of how property rights come into being.[1] This literature has generally recognized that the production of property rights is an economic activity determined by the expected returns to defining and enforcing property rights. The determinants of these returns include such variables as the value of the resource in question, homogeneity of the population competing for the resource, and the technology available for defining and enforcing property rights.

There have only been a few studies of technology's impact on property rights definition and enforcement and even fewer studies of the impact of institutions on the development of property rights technology. This volume helps to fill this void, with eight chapters discussing both the influence of technology on property rights and the effect of institutions on technological change as it relates to property rights.[2]

Before considering the specific examples, it is useful to place them into a theoretical framework.[3] In 1975 we hypothesized that the degree of property rights definition and enforcement activity is determined by the marginal benefits and marginal costs of devoting effort to definition and enforcement activity. All other conditions being equal, greater definition and enforcement activity will lead to clearer specification and protection of rights. For example, additional strands of wire added to a fence will increase the likelihood that what is inside the fence will stay in and that what is outside will stay out, but the additional value will come at a diminishing rate, i.e., the sixth strand will add value but not as much value as the fifth strand. On the cost side, adding strands will have rising marginal costs. The combination of diminishing marginal value and rising marginal costs places a limit on how many strands of wire is optimal and hence on the level of definition and enforcement activity.

IMPACTS OF TECHNOLOGY ON PROPERTY RIGHTS

The level of activity will change as the benefits and costs shift, and technology can affect the degree to which property rights are defined and enforced in three basic ways. First and most obviously, technology can change the costs of establishing property rights. If it lowers the cost of definition and enforcement activity, more activity will result and vice versa. A prime example of this cost-lowering change was the invention of barbed wire in the American West (Anderson and Hill 1975). Because of the aridity of the Great Plains, large expanses of land were required to support livestock. Thus, economically sized units were much larger than in higher rainfall areas, but appropriate fencing materials were also scarcer. This meant that it was expensive to fence livestock on a large scale and, consequently, most cattle ran on the open range. The need to better define property rights in land and to prevent livestock from wandering, however, created incentives for the development of new technology. In three years, 1866, 1867, and 1868, 368 fence patents were issued (McCallum and McCallum 1965, 27). Joseph Glidden of DeKalb, Illinois, was the most successful barbed-wire inventor, and after his invention was patented in 1873, production increased rapidly.

Second, technology can alter relative prices of inputs and outputs, thus changing the incentive to define and enforce property rights. For example, the invention of a better plow could increase the value of land and hence the return on defining and enforcing property rights to prairie farmlands. Similarly, the extension of railroads across the United States made land more valuable, increasing the benefits of definition and enforcement activity.

The value of the resource can also fall as a result of technological change, as happened with the value of horses following the introduction of tractors. The real value of horses in Montana fell from $1,111.75 in 1918 to $280.66 in 1926.[4] This decrease in value made it unprofitable to maintain property rights in all horses, and ranchers turned many horses loose on the open range, creating large herds of unclaimed animals.

Third, definition and enforcement of property rights can be inextricably connected to production technology. For instance, before barbed wire, cattle were run on the open range and cattle roundups were necessary for branding and marketing cattle. Because of economies of scale, it was not cost effective for a single cattle company to round up, brand, and market its cattle on the open range. Thus, numerous cattle companies would cooperate in open-range roundups, and these cooperatives ultimately provided the basis for cattlemen's associations. Not only did the roundup allow realization of scale economies, it also facilitated excluding outsiders from the range. Michael Malone and Richard Roeder (1976, 123–24) reported that "if an interloper tried to crowd his stock onto someone else's range, local ranchers could refuse to allow him the privileges of belonging to the roundup district." One entrant who had not secured permission to access the open range was "condemned . . . for this violation of range law and warned . . . that they would not handle his stock or cooperate with him in any way." Put simply, if an outsider could not participate in the roundup, he could not make a profit.

THE IMPACT OF INSTITUTIONS ON TECHNOLOGY

Just as technology can affect the formation of property rights, the institutional environment in which property rights evolve can also influence the development of technology. Obviously, security of returns to investing in technological change provides more incentive to invest, but the process of technological change has remained largely inside a black box.[5] Even less well explored is how investments in property rights technology will be endogenously related to the legal framework in which property rights evolve.

It is not our intent to develop a complete theory of this relationship, but we can be sure that the propensity of the formal and informal legal system to recognize property rights will affect the formation of technology to define and enforce those rights. For example, to the extent that common and statute law encourage the evolution of property rights, the more people will invest in technology to define and enforce their property. If common law focuses more on who has what rights and who is violating those rights, we can expect the development of more technology to measure harm. Similarly, common law application of injunctive relief, as opposed to compensatory relief, can have implications for technological investment. The former requires that a defendant cease and desist an activity until he can show that there is no harm, whereas the latter allows the harm to continue until compensation is paid.

WILL PRIVATE PROPERTY RIGHTS BE FORTHCOMING?

Even if the benefits of defining and enforcing property rights rise relative to the cost, there is no guarantee that more effort will result in the formation of private property rights. As Gary Libecap (1989, 19–28) has pointed out, there are variables that influence whether individuals and groups will be successful at "contracting for property rights." Two of the more important are the size and transparency of the gains from establishing property rights and the distribution of those gains.

If the benefits from clearer specification of property rights are large and visible, there will be more incentive for people to invest in defining and enforcing property rights. For instance, in Colorado, water in the Denver area has become much more valuable because of population growth. This has led to more definition and enforcement efforts through careful measurement of water flows. This helps explain why much more expensive and sophisticated technology is used today compared with the earlier days on the frontier.

But sometimes the size and transparency of benefits has not been sufficient to overcome the institutional inertia for regulation and to lead to property rights solutions to the tragedy of the commons. Pressure for individual transferable quotas (ITQs) has varied across fisheries depending on how obvious the overexploitation problem is.

The questions of who gains and who loses also affect the degree to which property rights evolve. Even if new technology creates substantial potential gains from new definition and enforcement mechanisms, those gains may not be distributed evenly among resource users. In such cases, the transaction costs of securing agreement among these users can be high, and new forms of property rights may not evolve. Johnson and Libecap (1982) argue that Gulf Coast shrimp fishers have been unable to escape the tragedy of the commons because of differential fishing abilities that allow the better fishers to do well even with open access.

THE ROAD AHEAD

Each of the chapters in this volume provides a mixture of theory and evidence regarding the interaction between technology and property rights. In chapter 1, Bruce Yandle explains how property rights evolve to become either private property or regulatory property. He also provides several case studies of the evolution of property rights in different institutional environments. Chapter 2, by Clay J. Landry, illustrates how technology has changed since the early days of mining and agriculture in the West to allow more precise definition and enforcement of rights and how this in turn allows more water marketing. In chapter 3, Robert B. Naeser and Mark Griffin Smith also discuss how sophisticated technologies are allowing efficiency-enhancing water marketing in Colorado. Chapters 2 and 3 illustrate how rising water values are leading to greater definition and enforcement activity. Barrett P. Walker's chapter provides an understanding of the potential for geographic information systems to more precisely identify the source and impact of storm runoff on watersheds and to charge polluters for their impacts. In chapter 5, Daniel Huppert and Gunnar Knapp examine how technological improvements lower the costs of defining and enforcing property rights to fisheries by making possible shipboard measurement of catch and identification of fishing boats operating in a particular area. In their chapter, Gregory B. Christainsen and Brian C. Gothberg discuss how factory ships have lowered whale-processing costs, thus increasing the value of the captured whale and worsening the tragedy of the commons in whaling. In chapter 7, Anna M. Michalak discusses the prospects and pitfalls of using contaminant source analysis to establish responsibility for air, water, or ground pollution. Her case studies suggest areas in which technology can better measure the cause and effect of contaminants, thus making common law remedies more available. David Gerard and Timothy J. LeCain, in chapter 8, show how a shift from injunctive relief to compensation under the common law has affected the incentive for firms to invest in pollution control technologies in the mining industry.

All these chapters give us a clearer understanding of the interface between technology and property rights and call for more intensive theoretical and empirical analysis of this interface.

NOTES

1. For a more complete discussion of the literature on the evolution of property rights, see Furubotn and Richter (1997, chs. 3 and 4).
2. The papers were prepared for a Political Economy Forum sponsored by PERC of Bozeman, Montana, December 2–5, 1999.
3. The theoretical structure of this introduction closely follows summary comments made by Gary Libecap at the forum. We are grateful for his insights.
4. These values are given in 2000 dollars using the CPI inflation calculator (http://146.142.4.24/cgi-bin/cpicalc.pl). The nominal values fell from $98 in 1918 to $29 in 1926. Technology was not the only influence. At the end of World War I, the U.S. Army stopped buying horses for its cavalry, and this also caused horse prices to fall.
5. For an early discussion of this issue, see Rosenberg (1982).

REFERENCES

Agricultural Economics Bureau. 1938. *Livestock on Farms, January 1, 1867–1935.* Washington, D.C.: U.S. Department of Agriculture.

Anderson, Terry L., and P. J. Hill. 1975. The Evolution of Property Rights: A Study of the American West. *Journal of Law & Economics* 18(1): 163–79.

Demsetz, Harold. 1967. Toward a Theory of Property Rights. *American Economic Review* 57: 347–59.

Furubotn, Erik G., and Rudolf Richter. 1997. *Institutions and Economic Theory: The Contribution of the New Institutional Economics.* Ann Arbor: University of Michigan Press.

Johnson, Ronald N., and Gary D. Libecap. 1982. Contracting Problems and Regulations: The Case of the Fishery. *American Economic Review* 72(5): 1005–22.

Libecap, Gary D. 1989. *Contracting for Property Rights.* Cambridge, England: Cambridge University Press.

Malone, Michael P., and Richard B. Roeder. 1976. *Montana: A History of Two Centuries.* Seattle: University of Washington Press.

McCallum, Henry D., and Frances T. McCallum. 1965. *The Wire that Fenced the West.* Norman: University of Oklahoma Press, Publishing Division of the university.

Osgood, Ernest Staples. 1929. *The Day of the Cattleman.* Chicago: University of Chicago Press.

Rosenberg, Nathan. 1982. *Inside the Black Box: Technology and Economics.* Cambridge, England: Cambridge University Press.

1

Legal Foundations for Evolving Property Rights Technologies

Bruce Yandle

Practically every normative examination of environmental problems by economists begins with a tragedy of the commons and ends with some proposed property rights solution. The property solution or rule may call for the definition and enforcement of common, public, or private property rights, but any escape from the tragedy requires a costly rationing mechanism. If the recommendation follows the Pigouvian tradition and calls for government taxation or regulation, the politically determined rules form a public or regulatory property regime (Yandle 1998). A Coasean recommendation relies on markets and a system of private property rights. It is important to note that any proposal for action implies some definition of property rights. Of course, enduring a tragedy and doing nothing can be the low-cost solution.

In describing the escape from the commons, stories about the evolution of property rights typically focus on demand thresholds where broadly described transaction costs are covered and entrepreneurs claim, define, trade, and are held accountable for certain rights to a former common-access resource (Anderson and Hill 1975; Barzel 1997; Cheung 1970; De Alessi 1998; Demsetz 1967; Libecap 1989; Ostrom 1990; Rose 1990). The process is thought to begin when the common-access resource is made more valuable by growing demand. But for the property rights threshold to be approached, cost-effective technologies for measuring, monitoring, and enclosing must emerge so that identifiable units of the resource can be claimed and transferred. Then the ultimate threshold crossing may occur provided the associated wealth distribution effects are successfully resolved (Libecap 1989, 11–12). Having sufficient gains from trade to cover market-related costs is necessary but not sufficient. Side payments may be required to satisfy those interested in maintaining the status quo. When property rights technology is lacking and the distributional cost hurdle is too high, private property rights may not emerge. The transaction cost wedge is simply too large. Political or regulatory property rights will emerge instead (Wiener 1999).

Among the environmental property rights storytellers, only those from the free market environmentalism school seem to focus on the importance of property rights technologies (Anderson and Hill 1975; Anderson and Leal 1991; De Alessi 1998; Libecap 1989). Among evolving property rights technologies, barbed wire presents the most dramatic efficiency story (Anderson and Hill 1975). Consider the salient parts of the barbed-wire story. With the introduction of barbed wire, it was possible to divide and enclose large amounts of grazing land at lower cost than occasioned by the alternative management scheme. Exclusive rights to vast sections of western territory were more readily enforced. As a result, cattle production was enhanced, land became more valuable, and demand for horses and cowboys declined.

Reference to the economist's reliable model of the competitive firm and industry seems to explain all this. There is growing demand for the capability to enclose and exclude. The foregone gains from trade are large. As demand grows and the cost of enforcing rights continues unaffected, the potential gain from reducing enforcement costs expands. Creative people observe the opportunity for profit. Inventions occur. Ergo, barbed wire is born.

All of this may sound reasonable and logical, but a critical element is omitted from the model, an institutional black box. The legal institutions that condition and affect the definition and enforcement of property rights need to be explicitly considered. For the barbed-wire story to hold, the cattlemen occupying western land must first be able to exercise the right to enclose the land. Libecap (1989, 60–64) tells us this was not always the case for western cattlemen; indeed their enthusiasm for barbed-wire enclosures led them to fence off illegally at least 36 million acres of U.S. government land.

For land to be enclosed and alienated and barbed wire sold, a technology must exist for measuring and recording land, as well as for measuring barbed wire. Then for the story to lead to full land-value maximization, the cattlemen must hold a recognized right to alienate the land. As residual claimants, owners of rights that can be defined, defended, and divested (3-D) to others, the holders of 3-D property rights have the incentive to search for and install barbed wire. If enforcement of these rights is lacking, the market for barbed wire will be severely limited; there will be little barbed wire other than in museums or in engineering laboratories.

The black box that regulates the property rights process contains competing providers of these institutional services. In terms of two extremes, the resource in question may be controlled and managed by government. Statute laws, regulation, and civil servants may form the dominant institution. Alternately, fully alienable rights to the resource may be held by private parties. Common law, contracts, and voluntary exchange may be dominant institutional ingredients. The development and implementation of property rights technologies will be significantly determined by which institutional provider is dominant.

This chapter explores the black box containing the legal institutions that accommodate and give legal force to property rights technologies. The chapter has three remaining parts. Starting with a theory from industrial organization, the next part builds a conceptual model of how property rights evolve. Various kinds of transaction costs that affect the evolutionary process are discussed, and property rights technologies address these. Private property rights technologies enter the evolutionary process, or perhaps enable it, at the point where alienability allows for private property rights to be distinguished from public and regulatory property. The phasing of property rights institutions matters in this. A different property rights technology may be required to support public property arrangements than for private property rights that are alienable. If public or regulatory property emerges first, then the incentives for the discovery of private property rights technology may be weakened.

There are two basic legal processes or competing institutions to consider as candidates for the legal institution component of the property rights black box. Statute law written by legislative bodies and common law that evolves through the courts can either accommodate technological changes that enhance market prospects or delay those changes. The discussion that follows provides theoretical and historical background on the evolution of environmental rights in common and statute law settings. The accommodating effects of new scientific knowledge and technologies are illustrated along the way. The chapter concludes with a discussion of certain aspects of the competition between providers of property rights institutions and a summary of key points developed in the chapter.

THE DIVISION OF PROPERTY IS LIMITED TO THE EXTENT OF THE MARKET

One of Adam Smith's insightful statements was made famous, perhaps, when George Stigler (1968, 129–41) treated the phrase as a theorem of industrial organization. Noting that the "division of labor is limited to the extent of the market," Stigler provides a solid theory of vertical integration and disintegration. Adam Smith's insight is used to explain how specialized activities within firms can become firms themselves and why they might not. The division of labor model is consistent with a demand threshold model of property rights evolution, what we might term a division of property rights model. Thinking of landownership as a bundle of potential rights, it is easily seen that lacking demand, some specialized rights might not be unbundled and traded in the market, just as some function within a firm awaiting growth in demand for a specialized service might remain integrated. For example, mineral rights might remain attached to the land package until the time that the expected value of the ore removed rises sufficiently. The right to discharge particular waste materials in a river might go unmarketed and in a way unrationed until the benefits of unbundling the specialized rights and trading these new property sticks exceeds the cost of building and operating the related institution.

If we reach back to the origin of property rights, we would likely find that all of the natural endowment was initially a commons. At some point it was economic to define, unbundle, and protect certain specialized parts of the commons as common property. Then, with additional increases in demand and encounters with scarcity, efficient behavior dictated further unbundling of specialized rights. Eventually the unbundling process led to the definition of private property rights for specific sticks in the property rights bundle that could be traded separately.

The notion of property rights specialization is captured in figure 1.1, which describes a highly stylized property rights flow diagram for one stick in a property rights bundle for a particular resource. The stick in question could be the right to the exclusive use of water flowing in a stream. The figure illustrates an evolutionary path of property rights formation that begins with a commons, moves to common property and then public property. At that point a fork is encountered where either fully defined, protected, and alienable private property rights emerge or some form of regulatory property is defined and maintained by government. The arrows in the figure indicate the general direction followed as economic and other social forces stimulate the evolutionary process.

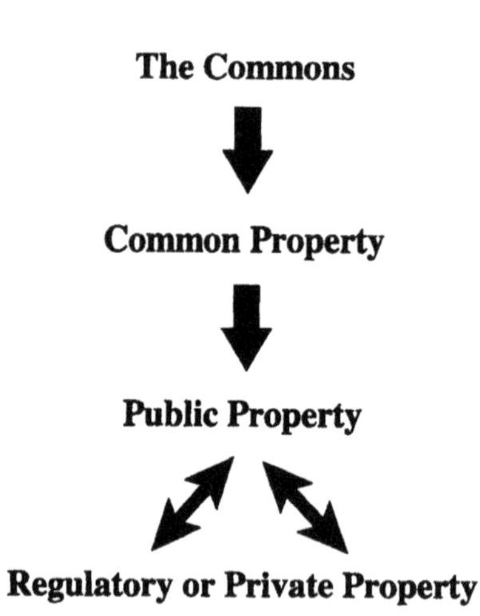

Figure 1.1. Evolving Property Rights

Little additional explanation is needed here for the first station in the flow diagram. A commons is simply a resource that has no coherent rationing mechanism. The law of capture determines use. In the case of water flowing through a pasture, there is no community enforcement mechanism; there are no rights. A shepherd hoping to water his sheep must rise early and move the flock ahead of a competing shepherd. When access is open, the sheep must drink deeply. Common property, the second station, leads to a first system of rights recognized by a larger community of people. A flowing stream under the control of a family, a clan, or club can be used or consumed by any member of the specified group. Consumption or use requires membership in the relevant group. Custom and tradition can serve to limit consumption within the group. The consumption rights cannot be transferred unilaterally by group members to nonmembers. The rights are defined and defended, but they may not be divested other than by common consent of the group.

Instead of pertaining to a family, clan, or club, public property implies the existence of a political mechanism—a government. Citizens have certain recognized rights to use and consume public property. Individual sticks or shares in the flowing water may not be claimed as the exclusive property of specific individuals. The public property rights are endowed as being common to the body politic; they are defined and defended but cannot be divested except by exercise of a political decision rule, such as majority voting. Rules and regulations developed by the body politic will serve to ration use.

Unbundling and 3-D Rights

As indicated here, the unbundling process is made up of three activities: defining, defending, and enabling rights to be transferred or divested—the characteristics of 3-D rights. These, in turn, contain elements of fixed and variable cost, with defining a particular stick in the property rights bundle the main fixed cost component. For example, land may be surveyed and the dimensions marked and recorded; water flows and interactions with pollution may be mapped using geographic information systems; stack emission monitors for measuring environmental use may be installed. In contrast, defending and transferring costs vary with threats and property rights activities. The interaction of fixed and variable costs generates a traditional set of short-run cost curves. There is an upward-sloping supply of property rights definition and defense. Changes in property rights technology can alter the mix of costs as well as relative magnitudes of the various cost generators.

History and tradition suggest that once a resource makes the transition from common access and becomes the exclusive property of a tribe, family, or political unit, the process is seldom totally reversed. So long as the once common-access resource has sufficient value to justify relatively low marginal cost border protection, the resource will at least remain in the public domain.

The formation of U.S. public lands from previously nonexclusive territory is evidence enough; however, the double-pointed arrows between common property and public property and between public and private or regulatory property indicate that property rights previously defined and enforced at a rather detailed level can and do return to common property or public domain status.[1] In the barbed-wire story, horses, which were a valuable property rights technology used in defining and defending property rights in pre-barbed-wire days, were displaced partly by wire, by tractors, and by the army cavalry. Many horses were released to the wild to become public property (Anderson and Hill 1975). In some cases, the horses and their progeny became common-access resources; they were not worth the cost of protecting them.

Movement from common property to private or regulatory property simultaneously creates and destroys rights, which causes distributional effects. When rights to a resource are held in common by an identifiable group, membership is a prerequisite for sharing the common resource. Access to fisheries, grazing range, hunting territories, even swimming areas are frequently managed by clublike organizations having specified membership requirements (Buchanan 1965). Membership in the property system may be tightly closed, and depending on the group decision rule, each member has some voice in determining the arrangement for sharing or transferring units of the resource (Holderness 1985). Even if direct voice is lacking, each member of the group can have well-founded expectations for sharing in the benefits of the common property. Of course, the group's identity can be so elastic that what first appears to be a closed system is really open for all practical purposes. For example, sharing arrangements to tribal lands may be available to progeny of Native Americans with particular tribal affiliations. Then, the definition of bloodlines and ancestry can be so elastic that potential property claims are practically unlimited. The accompanying uncertainty chills all private incentives to improve the land.

All else equal, the rising transaction costs that accompany the need to obtain agreement from an expanded or open-ended set of claimants diminish the prospects for gains from trade. Alienability and openness are inversely related. Fewer transactions will be observed in more open groups. If a private association or club formed for managing common property expands its membership beyond the limits that provide expected gains on average for each member, it is possible for a move to public property to be viewed by existing club members as wealth increasing. They may favor public sector management, hoping to obtain a meaningful output restriction or larger share at lower cost. Government and sovereignty may be invented, which introduces public property. At the same time, movement from common to public property can dilute the value of rights (or voice) held by a smaller, defined group within the polity while expanding group membership to include all citizens. There are distributional tradeoffs that follow from this.

Consider a person who is not a member of the common property group. If citizenship can be obtained at lower cost than group membership, then movement to public property opens the system for that person, along with all other citizens. Rights (or voice) are created for that person as he participates in the shared use of the resource, but shares are diminished for each member of the former common property group. We observe this with the demise of tribal fishing rights for Native Americans in the Pacific Northwest that accompanied the arrival of the new American citizens (Higgs 1982). Within-group contracting costs are relatively low among the members of each competing group, but are extraordinarily high for transactions that involve parties from different groups. For all parties taken together, new and old, the expected transaction costs for contracting increases. Might makes right; the previous Native American market process suffers. The incentive to conserve the resource is replaced by an incentive to exploit the resource. Enforcement costs rise with openness.

A transition from public to private property rights by definition produces a closed system (Holderness 1985). All rights are allocated; to participate in the use of the resource, an individual must obtain rights from some private right holder. There is one-to-one mapping between the number of rights and right holders. If the rights in question are fully alienable, any change in ownership requires unanimous agreement, which is to say that any individual owner has unilateral powers to transfer his rights to another party. By comparison, movement from public property to regulatory property along the other path partially reopens the system, relative to private property, but does so by political means; to use the resource, the qualifying individual must be favored politically and satisfy regulatory requirements that ration use. The rule of unanimity is generally broken; voting rules of less than unanimity and competition among special interest groups for political favors affect outcomes. Taking the full process into consideration, figure 1.1 implies that as property rights evolve, the degree of openness first contracts and then expands relatively, then contracts again. In the process, wealth is created, destroyed, and redistributed. Different types of property rights technology emerge along the way.

A few points of clarification about the highly stylized flowchart are in order here. Property rights evolution may be short-circuited, as when an immediate move is made from a commons to private property rights. Individuals have historically entered the wilderness and made private property claims that were enforced by the community or state. In a somewhat similar way, property rights that were at one time regulatory can become a commons. For example, operating certificates once required and issued by the Interstate Commerce Commission for providing transportation services are no longer needed. Deregulation of surface transportation has destroyed the regulatory property and converted the market to a commons.

Transaction Costs: The Achilles' Heel of the System

The discussion of evolving property rights and open versus closed systems has emphasized one part of the transaction cost problem, the costs associated with gaining agreement from an ever-expanding group of rights holders. There is another element of transaction costs that is of key importance to this paper—the cost of measuring and monitoring the items transferred among trading parties. Alienability and gains from trade cannot be obtained and contracts cannot be enforced where the ability to define, defend, and monitor is lacking. A bundle of private property rights can exist, making the system closed, but no transfers of certain sticks in the bundle will be observed. Examples of this include rights to instream flows in some locations; riparian and some mineral rights that accompany landownership; certain hunting, grazing, and fishing rights that cannot be transferred from the certified individual; and a multitude of informal arrangements for sharing common property (Ostrom 1990). In these cases, the desired right, which cannot be unbundled, can be obtained by acquiring the related land. Alternately, defining and defending may be feasible, but rising transaction costs associated with gaining agreement to move from an open to a closed system may eliminate the prospects for individual ownership and trade. It is possible to measure and monitor the withdrawal of oil from Alaskan public lands, but gaining agreement on changes in the terms and conditions for withdrawal is prohibitively costly.

Defining, defending, and gaining agreement are processes that depend on three categories of property rights technology. Defining and defending touch two categories. For any type of property right to exist, the resource itself must be identified and its amount and quality measured: This is category one. To defend the defined rights to the resource, threats and harms must be identified and measured: category two. These category two activities also accommodate rationing and exchange, which requires resource activity to be monitored and recorded. These categories of technology can include basic scientific knowledge about the resource and threats to it. The technology may include meters, remote sensing and recording devices, fences, brands, and identifying marks. The third category, which is important to gaining agreement, generally involves the use of some kind of two-way communication technology that provides information, feedback, and recorded agreement or disagreement between and among people who would normally transact with each other. These technologies can be as simple and subtle as council meetings in which senior community members discuss issues and reach agreement on resource use or a telephone call between a broker and client during which an agreement is made and a contract formed. They can be as complex and costly as satellite sensors that emit signals to interested parties who, viewing the same data, make decisions in a virtual council meeting. Among the technologies are some that are best suited for private property protection and others that satisfy special regulatory requirements. In other words, there is a technology fork that is the counterpart of the property rights fork in figure 1.1 in which property rights become either private or regulatory.

The Fork in the Property Rights Path

The private/regulatory property fork in the path of property rights evolution shown in figure 1.1 is the focal point of any discussion of evolving rights for environmental goods. This junction point is also the location where incentives for the development and use of different categories of property rights technologies live or die. One might assert with little objection that government regulation of U.S. environmental goods is currently so pronounced that there is hardly any activity along the private property fork in the road. Is this the case because of technical inability to measure, monitor, and enclose? Or is the public sector dominant because of opportunities provided politicians to redistribute wealth? There is evidence to support both cases. Quite possibly, the absence of property rights technology provides the initial nudge along the regulatory property route. Then, rent seeking and redistribution take over, making it costly to deregulate and privatize. Holding the potential gains from trade constant, once regulatory property is formed, there are few if any incentives to develop private property rights technologies. We have a path-dependent story (North 1990, 94). There are separate paths that generate path-specialized property rights technologies.

We can assign cost in terms of foregone gains from trade to the degree of divergence between the regulatory and private property paths. Cost may be highest when regulation totally precludes the ability of resource managers to make efficiency-enhancing adjustments in the use and deployment of resources. That cost falls as managers are allowed some flexibility.

Consider the U.S. pollution control experience. The property rights junction became meaningful in 1970 for air quality and in 1972 for water quality with the passage of fundamental federal statutes. The two statutes effectively nationalized the two environmental assets, embraced a technology-based command-and-control regime, and ended a period of property rights history that relied significantly on decentralized regulation, private property rights, and market forces. In a sense, those early statutes recognized the magnitude of the regulatory task by using technology as the monitoring mechanism. In effect, the new landlord, the federal government, indicated that it would not measure and monitor environmental quality directly. Instead, the landlord established technology-based standards for broad categories of production processes within specific industries.[2] If the standards were met, permits were issued. From that point, the regulator would check each user and determine if the user was properly equipped. If so, it was assumed that the outcome would be in the interest of the landlord. Since, by statute, each and every discharge source had to be controlled by specified technologies, there was no possibility for trade, even within the confines of a particular plant. The incentive to discover property rights technologies was practically not existent.

Technology-based command-and-control regulation generates regulatory property rights in the form of a nontransferable government-issued permit. Defining and defending the rights is the business of the government, assisted by regulated firms, environmental organizations, and interested citizens. Under the command-and-control regime, outcomes and environmental conditions play second fiddle to inputs and the appropriate use of technologies. With trade forbidden, there is no market incentive to discover and apply new techniques for monitoring, measuring, and legally packaging units of pollution rights. There is an incentive to meet the technology demands of the regulation and to impose those demands on competitors.

Whether operating in a common law or regulatory regime, efficiency dictates that economic agents must be allowed to truck and barter as they juggle use of environmental resources. Sometimes, use of scarce environmental assets requires ownership of particular land parcels, membership in some specified community or tribe, or citizenship in some political unit. At other times, access and use is simply a matter of engaging in mutually beneficial exchange of fee simple rights with rights holders. The degree to which property rights can be alienated determines the wealth creation potential for the resource in question. The limits of efficiency are found when defined and defended environmental rights can be completely alienated.

Holding constant the demand for a bundle of environmental rights, the supply of activities that define, defend, and divest various sticks in the bundle or the entire bundle depends on the technology of property rights definition and enforcement. Wealth depends on identification of ownership when rights are transferred among trading partners. When the ability to measure and subdivide parcels from a larger tract of land is lacking, the asset owners cannot maximize the potential value of the land. If water flows cannot be metered and water users sent bills, then water will not be provided by private suppliers. If particular emitters of harmful pollutants cannot be distinguished from a host of similarly situated potential emitters of the same pollutant, then the scope for bargaining and use of rules of liability will be reduced. Command and control will rule the day. Evolving property rights technologies enable movement along the path from the commons to private property.

BUILDING LEGAL FENCES THAT DEFINE AND DEFEND ENVIRONMENTAL RIGHTS

At the time of the barbed-wire story, there were two basic forms of law in the United States. There was common law, also referred to as evolved and private law, and statute or administrative law. These more formal legal systems augmented informal customs and traditions. The same is true today. The common law emerges on a case-by-case basis from real controversies adjudicated by common law judges (Hayek 1973; Leoni 1991; Meiners 1995; Yandle 1998; Zywicki 1999). Common law evolves in a small-numbers setting. Through judges' traditional use of precedents in deciding cases, the law is generalized to a large number. Statute law emerges in a large-numbers setting, but can be specialized to apply in selected ways to smaller numbers of people and technically specific situations. The common law process is continuous; an opportunity for modification and the introduction of new knowledge is afforded each time a common law judge writes an opinion. By contrast, the statute law process is continual. Statutes are written when the legislative body is in session and when the political agenda calls for action. Any person who writes a contract has written common law. Only elected politicians can write statute laws.

Now consider the market entry of barbed wire under two distinct situations. Situation one: If legal prohibitions are absent, and if common law judges will enforce property lines defined by barbed wire, as well as by other methods, then cattlemen and barbed-wire producers have an incentive to pursue gains from trade. The market process is fully competitive. Situation two: If the legislative body by statute outlaws the use of barbed wire, all bets are off. Or if the legislative body establishes a land management commission with delegated authority to define and regulate the use of appropriate technologies for enclosing land, then another set of incentives enters the picture. In the first situation, resources are expended in a competitive process as barbed-wire sellers seek to gain the patronage of interested cattlemen. In the second situation, inventive entrepreneurs must leap two hurdles—the invention, production, and marketing of barbed wire and successful lobbying to obtain regulatory approval of the use of barbed wire. Indeed, the producers, or just one producer, may lobby successfully to have barbed wire or one patented type of wire specified as the only state-approved method for defining property rights.

The two market entry situations generate different competitive and rent-seeking responses from barbed-wire and other enclosure system producers. Barbed wire may become an approved enclosure device by way of custom, contract, or statute (Ellickson 1994). There is obviously no low-cost way for particular barbed-wire producers or the industry to effect a change in custom or to influence the content of all land-use contracts.

There is also no low-cost way for the special interest group simultaneously to obtain favorable outcomes from hundreds of common law judges. In contrast, the legislature and regulators provide low-cost opportunities for favor seeking.

Evolving Property Rules for Quantity and Quality

Scientific discoveries obviously play a crucial role in the definition and protection of private property rights and forming markets for environmental rights. Consider water quantity and water quality, two distinctly different commodity characteristics. If we start at the grossest level, the flow of water through an irrigation ditch, we can imagine the relative ease with which ordinary people can stand at an irrigation ditch and observe when and where water is flowing. Measuring the rate of flow and total volume in a given time period is obviously more complex. For water quantity, we should expect that crude rules based on first occupancy would emerge first. Then, appropriative rights with a "use it or lose it" condition might come next. Later, with low-cost time and place metering, we would expect to see a more sophisticated rule of law, one that accommodates water transfers and markets.

By comparison, developing institutions for managing water quality is far more daunting, if for no other reason than the scope of qualitative dimensions that might be rationed. Having identified the scarce quality dimensions to be rationed is just the beginning. Sources of degradation and use must be identified along with the loss and losers of benefits associated with particular levels of water quality. Consider the problem of ensuring the healthfulness of water for human consumption. We can see that the capability of certifying healthfulness would make it possible to expand the market for drinking water. Other more costly ways of ensuring quality would prevail if the ability to measure and certify were lacking. Consider the use of rivers for disposing of waste. Being able to measure levels of dissolved oxygen makes it possible to set related constraints and market consumption rights. Each characteristic in the water quality bundle that might be scarce and valuable comes with a measurement challenge. Being able to measure and meter does not necessarily solve the control problem: The ability to measure the level of dissolved oxygen in a stream does not translate into technology for reducing the consumption of suspended oxygen. Without the ability to meter and monitor individual consumption of water quality, it would be impossible to know when property rights were held, transferred, or taken. In the absence of scientific knowledge, it would be impossible to move beyond the stage where the mysterious nature of the environment causes people to avoid risky encounters with environmental use.

How U.S. Water Rights Evolved

In her rich discussion of evolving U.S. water law, Carol M. Rose (1990) describes what at first seems to be a property rights puzzle: An evolved system of private property rights seems to decay and become common or public property rights. In the earlier years of industrial development in England and the eastern United States, water in streams was valuable primarily for providing motive power to mills. The issue was water quantity, and the rules of law and property were rather clear. First, the property rights of ancient uses of streams were given priority at common law. After that came the rule of first occupancy (Rose 1990, 273). Generally speaking, first occupancy meant the first owner had to make a water-dependent capital investment. Newcomers to streams understood that they would have to bargain for rights held by established users. The world was Coasean.[3] The appropriative rights system in the western United States was similarly well grounded. In both cases, measuring and monitoring use was relatively easy. An impartial observer could quickly determine if a stream was diverted away from a waterwheel or if an irrigation ditch was opened or closed. No special legal machinery was needed to accommodate trades.

Rose's puzzle arises when accelerated economic development in the eastern United States led to a revision in water law. Leaving the relatively well-founded rule of first occupancy, the law moved not to a rule of prior appropriation, like that which developed in the arid West, but to a seemingly softer property rights doctrine of reasonable use within a riparian system. What had previously looked a lot like private property rights became common property in a multiple-user setting. But as Rose suggests, what appears to be a puzzle is really no puzzle at all. Accelerated economic development delivered the need to ration simultaneously two attributes of water—physical supply and the quality of the water supplied.[4] Physical volume could be monitored and metered at low cost. Markets for rights to water flows came with relative ease. Water quality was another matter entirely because each user's activity affected the level of quality available to all future users. Use determined quality.

When identifiable mines or farms consume an acre-foot of water, water quantity is a private good. When an acre-foot of water receives waste discharge from multiple users, the resulting water quality is a collective good. Lacking private ownership of instream water quality and lacking the technology to identify water quality consumption, reasonable use became the rule of law. Rights to water flow were common property held by any and all riparians. Each user was entitled to make reasonable use of the flowing asset. The large number of users and associated transaction costs were the culprits that limited the use of a rule of prior appropriation and markets.

This account of evolving water law, technology, and rules of property tells us about demand for two fundamental traits of water supply—water quantity and water quality. The technology for rationing quantity is far simpler than that for quality, which itself is multidimensional. When water use involves common pools and large numbers of simultaneous users, the definition and enforcement of rights becomes even more complicated. The evolving pattern of law that dealt with these issues is described in table 1.1. When the number of water quantity users is small, individualized private property rights seem to emerge without difficulty. When the number of users is large and the users draw from a common pool, common property with rules for sharing emerges. Community water systems with membership and user fees arise. When water quality is the trait under consideration, common property seems to follow, no matter the size of the user group. River basin associations and other firmlike approaches emerge to "own" and manage the common property (Riggs and Yandle 1997).

Underlying the data in table 1.1 are issues involving technologies. Recalling the earlier technology analysis, there are three categories of property rights technology involved in defining, defending, and divesting property rights. Two categories, the technology for measuring and monitoring use and technologies that reduce the transaction cost of communicating and gaining agreement, are especially relevant to the issue of group size. If low-cost technologies are available, then private property rights may emerge for quality and quantity irrespective of group size. This assumes that the technology for solving the water quality problem is available. But as table 1.1 suggests, the availability of low-cost technologies that identify individual consumption of water quality and quantity still seems to be a limiting factor for managing water quantity among a large number of users.

If the pattern of costs imposed by particular dischargers can be measured and monitored at low cost, then private rights and markets can evolve in the small numbers and large numbers case. A focused review of some court decisions and regulatory actions will illustrate how science and technology assisted these changes.

How Science Reduced the Costs of Defining and Defending

As the pace of eastern industrial development quickened and the locations for water-driven mills became filled, pollution became a more serious problem than water supply. Theoretically, riparian rights would protect downstream occupiers of property from having their rights degraded by upstream

Table 1.1. Property Rule Variation

	Small Numbers	*Large Numbers*
Water quantity	Private rights	Common property
Water quality	Common property	Common property

dischargers. If an upstream discharger imposed cost on a downstream party, the damaged party had a cause of action. The remedy was the granting of damages and injunctive relief. To be successful in court, the plaintiff had to show evidence of harm caused by the defendant. The common law acted as a legal fence around the property rights of downstream landowners. The burden of building a physical fence to protect riparian right holders was placed on the upstream discharger. The polluter could either develop a technological fix, contract around the rule, or violate the property rights and face the consequences. The economic calculus involved in selecting the logical option was heavily influenced by scientific knowledge about pollution and harm, its source, how to measure it, and how to fence it in.

Meiners, Thomas, and Yandle (1999) tell how the discovery of germ theory in the 1890s led to the rise of an early U.S. environmental movement calling for the use of science in managing water quality. Before this breakthrough, there was vague recognition that human health was related to water quality, but the exact linkages were not understood. With germ theory established, the Massachusetts Board of Health soon discovered the relationship between sewage in rivers and typhoid. Then, enhanced filtration technology was developed that eliminated the germs (Thomas 1999, 15); this made it possible for sewage producers to fence off property rights. But at the time germ theory was developing, industrial firms were shielded from common law suits by newly developing state regulation. They were enjoying regulatory property rights.

The typical state regulatory body wrote technology-based standards for firms that discharged in rivers and streams; these were applied to new sources. Industry in place continued to operate unabated. There was little that downstream landowners could do. In the absence of scientific evidence of harm, downstream receivers of waste could not have their day in court. Existing polluters enjoyed preemptive rights that could not be challenged. But the new scientific knowledge emboldened citizens in protecting their common law rights; using germ theory they were able to provide evidence of harmful invasion of their riparian rights. And the new filtering technology lowered the cost for common law judges to enforce a strict property rule. With common law rights reestablished for downstream parties, water quality users became members of a closed system. To engage in activities that deteriorated the water, a discharger had to first negotiate with downstream holders of property rights. Scientific knowledge about the linkage between waste, germs, and disease set in motion economic forces that led to the development of improved pollution control machinery and the reinforcement of riparian property rights.

With the rise of the U.S. industrial economy, scientific knowledge of the linkage between human health and pollution combined with common law actions to make a compelling case for protection of environmental rights. By 1917, a number of industrial states—Connecticut, Ohio, and Pennsylvania—had state boards for controlling industrial wastes. By 1923, the American Water Works Association (1923) could report that industrial pollution had damaged some 248 water supplies in the United States and Canada. Phenol discharge from coke works topped the list of harmful pollutants. With the science established, and faced with pressure being exerted by state health boards and the real threat of science-based common law suits, firms along the Ohio River joined in a voluntary effort to reduce phenol discharge into the Ohio River. Technology for removing phenol from wastewater was discovered and installed. By 1929, practically every phenol discharger had elimination devices at work. Fear of science-based common law suits partly motivated the cleanup.

Improvements in the science of hydrology made it possible for common law judges to identify polluters who poached on the property rights of downstream owners and led to major modification in common law rules involving groundwater pollution. In *Wood v. Picillo* (1982), neighbors brought suit against a farmer whose waste dump had allegedly contaminated their drinking water. In holding for the plaintiffs, the Rhode Island Supreme Court overturned a 1934 precedent that would have favored the defendant. Noting improvements in the scientific understanding of groundwater that had developed since 1934, the court indicated that groundwater contaminants could now be traced to their source, making it possible to provide strong property rights protection to downstream parties.

Property Rights Technology and Community Action

These cases involve a small number of parties and thereby provide the better possibility for private property rights. Transaction costs are relatively small, and the technical challenge posed by identifying cause, effect, and damages presents a low barrier for defining and defending private property rights. When the number of interested parties is large and the environmental outcome is collectively determined, the property rights technology problem becomes more complex and costly to resolve. When that is the case, history tells us that communities organize around the resource in question, form a transaction cost-minimizing virtual firm and treat the environmental resource as a managed asset. Then the management of the firm has the responsibility of defining, defending, and divesting property rights. Three well-known water quality stories illustrate the point.

Heavy pollution in the Ruhr, Emscher, and Wupper Rivers and serious human health consequences in nineteenth-century Prussia led government to define, allocate, and defend public property rights in water quality for these three river basins (Bower 1981). Today the water management organizations act with ownerlike concern as they build treatment plants, determine and collect fees, manage parks and recreational areas, and produce and sell drinking water. To the extent possible, the public sector managers charge prices and manage environmental assets as though they were private property.

Prior to building a facility that discharges into these waters, the plant owner must provide scientific data on the effects of the discharge on water quality to the river basin association managers. Then the river basin managers take samples of the discharge and run ecological tests using flora and fauna from the rivers, providing the river basin associations with geographic information systems that enable them to model the effects of discharge on water quality at different points in the system. Discharge fees are assessed on the basis of this technical information.

What might be the next move in property rights evolution, the formation of fully alienable private rights to consume specified units of water quality, has not taken place, nor is it likely to. The prospect is not bright for further efficiency gains that could come when dischargers and environmentalists buy and sell rights in open markets. Indeed, I was told on my last visit to the Ruhr headquarters in Essen that there would never be another river basin association in all of Germany.[5] In saying this, my host, who was not an economist, was reflecting on the politics of the situation. The existing associations are products of another era, when it was difficult for special interest politics to overturn the decisions or a king. Today Germany is a leading participant in the European Union (EU). There is a new mix of political and economic forces at play; decentralized river basin associations are not a part of the current mix. Even so, when the gains from trade are large enough to fund the distributional effects, we predict an increase in German and other European river basin associations.

At this point the German associations are a partially closed property rights system. The associations "own" certain specified rights to rivers, manage the rights, and sell access to them. The associations maintain water quality and quantity constraints in which there is no apparent political meddling. By contrast, the other rivers in Germany are managed as open systems and are controlled by statutes and regulation. Environmental organizations and other special interest groups have voice in determining outcomes, and polluters hold regulatory property rights that enable them to limit competitive entry. Property rights technology has emerged in the association-run water quality systems but not in other systems.

A similar story can be told about the Ohio River and the Ohio River Sanitation Commission (ORSANCO), the river association that resulted from a 1948 multistate compact for the purpose of restoring water quality to safe levels (Cleary 1967; Maloney and Yandle 1983). In fact, the Ohio and Ruhr Rivers share similar histories. Both were major industrial rivers located in coal and steel territories, and experienced serious degradation from the discharge of human waste. The stimulus that shifted property rights for the

Ohio came when microbe carriers of gastroenteritis traveled upstream from Cincinnati and imposed cost on Pittsburgh and other upstream cities that had routinely discharged untreated human waste into the river. In an interesting way, Cincinnati got revenge. With the compact in place, the Ohio River ceased being a commons and became subject to a system of public property rights managed by ORSANCO. Given numerous dischargers along the path of the Ohio, water quality at any given point was a collective result. To bring transaction costs to manageable proportions, the river had to be organized as a firm, and the firm had to develop the technology to define and defend its asset.

ORSANCO did this by developing a system of continuous water quality monitoring with data transmitted to its Cincinnati headquarters. The new technology that could scan and measure a number of individual water quality parameters reduced the unit testing cost to 7.5 cents (Cleary 1967, 210); the older laboratory technique cost $2.20 per test. With monitoring costs reduced, the association then imposed custom-tailored cleanup standards on dischargers in the basin. Water quality and public health improved. The ORSANCO primitive water quality monitoring system served as barbed wire in protecting water quality in the Ohio River.

Marketlike Regulatory Property Rights

At times, it seems, prospects for gains from trade become so large that marketlike regulatory approaches emerge, evolving to something closer to private property rights. The U.S. experience with its regulation-induced air pollution permit markets provides a situation in which the regulatory property rights technology could be transferred to a private property rights regime, if there were demand for the resources independent of the regulation itself.

In the development of federal air pollution control, command-and-control regulation supplemented and supplanted common law property rights protection. At common law, a damaged owner or occupier of land or an entire community could bring a nuisance suit against a polluter or negotiate with the polluter and receive compensation for damages. The ability to define and defend property rights was driven by market forces and the law of private property rights. By contrast, a marketable permit regime imposes an emission standard to be met and requires polluters to show that their or someone else's equivalent emissions are reduced by a stipulated amount. Once the collective constraint is met, individual dischargers can generate a tradable asset by reducing emissions beyond the amount stipulated. But although the regulatory property rights in the form of marketable emission allowances facilitate trade among certified members of the regulated community, the value of the permits is determined strictly by the opportunity cost of pollution control, which is determined itself by government-imposed specification standards. It is regulation and only regulation that causes the right to emit one ton of sulfur dioxide or a milligram of phosphate loading to have value. Regulation specifies the level of scarcity and determines the potential for gains from trade. The utilitarian basis for the value is uncertain at best.

California's southern coast, where according to the South Coast Air Quality Management District (2000) 15 million people breathe some of the dirtiest air in the United States, offers the most prominent example of the breakout of air pollution control markets. A rather substantial set of market institutions is now in place. Under California law, participants in marketable permit markets must have continuous emission monitoring capability and report emissions data that can be audited. Then, if a firm reduces emissions by more than the required amount, the firm can record and bank its regulatory gains with the authorities and receive a deed for the reductions; the deed is a transferable asset (Touchstone Environmental 1999, IIA, 2.6). California's regulatory property rights system has generated private property rights technologies for managing air quality.

FINAL THOUGHTS ON COMPETING PROPERTY RIGHTS PROVIDERS

This analysis of how competing legal processes can accommodate and encourage flows of private property rights technology has traveled a long journey that began with a commons and ended with private property rights. Along the way a theory about the division of property rights was developed and applied to environmental resources. Building on the demand threshold model used by other property rights scholars, the model developed in this paper emphasizes the development of property rights technology as a transaction cost-reducing mechanism. Tracing the development of property rights to U.S. environmental resources through common law and regulation shows that the gap between regulatory and private property rights enlarges and contracts as technologies enter and overwhelm rent-seeking forces that stand in the way of the development of new markets for environmental resources. This analysis provides certain insights about property rights.

- Common law embodies few incentives for rent seeking and creates more incentives for technological change that lowers transaction costs of defining, defending, and divesting sticks in the property rights bundle.
- The fork in the road between regulatory property and private property also embodies different incentives for technological change.

- When private property rights are too expensive to define and defend, virtual firms or community-based organizations can evolve that can solve some of the transaction cost problems.
- Both virtual firms and regulatory solutions are more likely to freeze in place particular technologies than a regime of private property rights under common law.
- Regulatory property rights will create inefficiencies, and as new technologies enter, the gap between regulatory and private property rights may become so large that quasi-property rights may evolve.

NOTES

1. The common property and public property distinction is ancient. Bracton, or Henry of Bratton, used this characterization of water rights in the mid-thirteenth century, as did another English treatise of unknown authorship referred to as Fleta. On this see Lauer (1963).

2. There was a "one suit fits all" aspect of this that imposed costs on all but the "average" production process. The obvious economic hardship was the basis of *E. I. du Pont de Nemours and Company v. Train* (1977), a water pollution control case arguing that permits should be individualized. Siding with the U.S. Environmental Protection Agency (EPA), the court argued on the basis of transaction costs, indicating that EPA had to deal with more 42,000 dischargers, which made it impossible for the agency to custom design permits and meet the statutory deadline. For discussion, see Findley and Farber (1995, 282–87).

3. It is interesting to note that an identical system of water rights developed in Japan in what is called Japanese Common Law. See Ramseyer (1989).

4. There is another issue as well. Exclusive riparian rights may not work effectively when there are interdependencies among users. The softer reasonable use doctrine turns out to be more efficient. Higgs (1982) gives details on how reasonable use maintained the economic value of salmon fisheries for Native Americans prior to the arrival of the Europeans, who destroyed the rights by using an appropriative rights property rights system.

5. Author's personal interviews, Ruhrverband headquarters, Essen, Germany, April 6, 1998.

REFERENCES

American Water Works Association. 1923. Progress Report of Industrial Wastes in Relation to Water Supply. *Journal of American Water Works Association* 10: 415–30.

Anderson, Terry L., and P. J. Hill. 1975. The Evolution of Property Rights: A Study of the American West. *Journal of Law & Economics* 18(1): 163–79.

Anderson, Terry L., and Donald R. Leal. 1991. *Free Market Environmentalism*. San Francisco: Pacific Research Institute for Public Policy and Westview Press.

Barzel, Yoram. 1997. *Economic Analysis of Property Rights*, 2nd ed. Cambridge, England: Cambridge University Press.

Bower, Blair T. 1981. *Incentives in Water Quality Management: France and Ruhr Area.* Washington, D.C.: Resources for the Future.

Buchanan, James M. 1965. An Economic Theory of Clubs. *Economica* (February): 1–14. Reprinted in *Readings in Microeconomic Theory,* 2nd ed., William Breit and Harold M. Hochman, eds. New York: Holt, Rinehart and Winston, 1971: 547–56.

Cheung, S. N. 1970. The Structure of a Contract and the Theory of a Non-Exclusive Resource. *Journal of Law & Economics* 13: 49–70.

Cleary, Robert. 1967. *The ORSANCO Story.* Baltimore: Resources for the Future.

De Alessi, Michael. 1998. *Fishing for Solutions.* London: Institute for Economic Affairs.

Demsetz, Harold. 1967. Toward a Theory of Property Rights. *American Economic Review* 57: 347–59.

Ellickson, Robert C. 1994. *Order without Law: How Neighbors Settle Disputes,* reprint edition. Cambridge, Mass.: Harvard University Press.

Findley, Roger W., and Daniel A. Farber. 1995. *Cases and Materials on Environmental Law,* 4th ed. St. Paul, Minn.: West Publishing.

Hayek, F. A. 1973. *Law, Legislation, and Liberty,* vol. 1. Chicago: University of Chicago Press.

Higgs, Robert. 1982. Legally Induced Technical Regress in the Washington Salmon Fishery. *Research in Economic History* 7: 55–86.

Holderness, Clifford G. 1985. A Legal Foundation for Exchange. *Journal of Legal Studies* 14(June): 321–44.

Lauer, T. E. 1963. The Common Law Background of the Riparian Doctrine. *Missouri Law Review* 28: 60–107.

Leoni, Bruno. 1991. *Freedom and the Law.* Indianapolis: Liberty Fund Press.

Libecap, Gary. 1989. *Contracting for Property Rights.* Cambridge, England: Cambridge University Press.

Maloney, M. T., and Bruce Yandle. 1983. Building Markets for Tradable Pollution Rights. In *Water Rights: Scarce Resource Allocation, Bureaucracy, and the Environment,* ed. Terry L. Anderson. San Francisco: Pacific Institute for Public Policy Research, 283–320.

Meiners, Roger E. 1995. Elements of Property Rights: The Common Law Alternative. In *Land Rights: The 1990s' Property Rights Rebellion,* ed. Bruce Yandle. Lanham, Md.: Rowman & Littlefield, 269–93.

Meiners, Roger E., Stacie Thomas, and Bruce Yandle. 1999. Burning Rivers, Common Law, and Institutional Choice for Water Quality. In *The Common Law and the Environment,* ed. Roger E. Meiners and Andrew P. Morriss. Lanham, Md.: Rowman & Littlefield, 54–85.

North, Douglass N. 1990. *Institutions, Institutional Change, and Economic Performance.* Cambridge, England: Cambridge University Press.

Ostrom, Elinor. 1990. *Governing the Commons: The Evolution of Institutions for Collective Action.* Cambridge, England: Cambridge University Press.

Ramseyer, J. Mark. 1989. Water Law in Imperial Japan: Public Goods, Private Claims, and Legal Convergence. *Journal of Legal Studies* 18(January): 51–78.

Riggs, David, and Bruce Yandle. 1997. Environmental Quality, Biological Envelopes and River Basin Markets for Water Quality. In *Water Marketing: The Next Generation,* ed. Terry Anderson and Peter. J. Hill. Lanham, Md.: Rowman & Littlefield, 147–66.

Rose, Carol M. 1990. Energy and Efficiency in the Realignment of Common-Law Water Rights. *Journal of Legal Studies* 19(June): 261–96.
South Coast Air Quality Management District. 2000. South Coast Air Quality Management District. October 27. Available: www.aqmd.gov. Cited: November 28, 2000.
Stigler, George J. 1968. *The Organization of Industry*. Homewood, IL: Richard D. Irwin.
Thomas, Stacie. 1999. Water Quality Management and Property Rights: A Financial Markets Test of Institutional Stability. Department of Economics, Clemson University, Clemson, S.C.
Touchstone Environmental, Inc. 1999. *The Complete Guide to Environmental Law*. North Vancouver, B.C.: Specialty Technical Publishers.
Wiener, Jonathan Baert. 1999. Global Environmental Regulation: Instrument Choice in Legal Context. *Yale Law Journal* 108(January): 677–800.
Yandle, Bruce. 1998. Coase, Pigou, and Environmental Rights. In *Who Owns the Environment?* ed. Peter J. Hill and Roger E. Meiners. Lanham, Md.: Rowman & Littlefield, 119–52.
Zywicki, Todd J. 1999. Environmental Externalities and Political Externalities: The Political Economy of Environmental Regulation and Reform. *Tulane Law Review* 73: 845–921.

CASES CITED

E. I. du Pont de Nemours and Company v. Train, 430 U.S. 112, 97 S.Ct. 965, 51 L.Ed.2d 204 (1977).
Wood v. Picillo, 433 A.2d 1244, S.Ct., R.I. (1982).

2

The Role of Geographic Information Systems in Water Rights Management

Clay J. Landry

In the western United States, water is allocated through a system of property rights known as the prior appropriation doctrine. In the early days of the doctrine, it was not uncommon for armed altercations to occur between water users in an effort to protect their water claims. To the chagrin of some rights holders, guns are no longer an acceptable form of protecting water claims.[1] Recently state water management agencies have taken on the role of defining and enforcing water rights.

The system to date has been largely self-regulated and driven by complaints, prompting state enforcement agents to investigate reports of unauthorized use. But water rights management is becoming increasingly complicated. With the introduction of new and varied uses for water,[2] water rights officials must reconcile the competing demands and needs of agricultural, industrial, municipal, and environmental rights.

Conflicts over water and reports of illegal use are on the rise. Inadequately defined water rights and high enforcement costs are two reasons illegal and unauthorized water use persists. Though states and individual water users find it simply too expensive to monitor and enforce the thousands of water rights on record, the growing number of complaints is driving increased efforts to improve water rights definition, enforcement, and allocation. Recent technological changes are assisting these efforts.

This chapter examines how state agencies and the private sector are using technology, in particular geographic information systems (GIS) and associated technologies, to manage water rights. Examples are used to illustrate how GIS, aerial photography, and satellite imagery can improve the accuracy and speed at which states can adjudicate water rights and settle disputed claims. The chapter discusses how GIS and associated technologies are reducing enforcement costs and allowing states to stop illegal water use. It also describes how these technologies are improving water rights management in the field and stimulating the development of water markets. In addition to these positive benefits, the paper examines controversial aspects of GIS and satellite imagery in the context of infringement on rights of privacy.

THE EVOLUTION OF WATER RIGHTS

The evolution of water rights in the American West has been a continuous process since settlement in the region began. Early pioneers brought with them the eastern system of riparian water law. This common law approach to water allocation gave the right to use water to those who owned property adjacent to streams, rivers, and lakes. The reasons settlers adopted riparian rights are not difficult to understand. First, there was no other competing legal system. Early judges and lawyers were familiar only with eastern law and thus transferred it to the legal system of the West (Anderson and Hill 1975). Also, land with adjacent water was abundant relative to the number of settlers—there was no reason to abandon the legal system as long as there were sufficient land and water resources. But as more people moved west, land with water became scarce. Water users soon realized the need for a new system of allocating water that accommodated the dry landscape, and the prior appropriation doctrine emerged as a competing water rights system.

Anderson and Hill developed a useful framework for understanding the transition and evolution of the prior appropriation doctrine. They use a marginal decision model to show that investment in property rights definition and enforcement is determined by the marginal benefits and costs of either of these activities. The model predicts that property rights activity increases as the market value or the probability of loss of a right increases. Similarly, the model predicts that property rights reorganization increases as the cost or resource requirements needed for definition and enforcement decrease. Changes in technology and resource endowments affect definition and enforcement activities (Anderson and Hill 1975, 167).

Throughout the West, the value of water increased because of new demands created by mining and agriculture. With it came an effort to reorganize water rights. The new water rights system grew primarily from the customs of gold mining camps. As surface deposits played out, miners abandoned their pans and shovels for more effective, but more water-intensive, hydraulic methods of mining. Water rights and ownership became important concerns as the demand increased for water to fuel high-pressure water guns and to wash gold flakes through sluice boxes. Miners had already adopted prior appropriation rules for establishing property rights to mining claims, so when the ownership of water became an issue, they applied the same approach.

Miners claimed water much like they staked a claim. Sometimes a notice was posted in a conspicuous place, usually at the point of an intended diversion, indicating the amount of water claimed. More commonly, however, they claimed ownership simply by diverting water. The construction of diversion canals served notice to other miners that a right was being claimed. Those who diverted water first obtained the priority or senior right. Subsequent appropriators were required to make do with what was left. Even if located upstream from a prior right, a junior user was required to allow enough water to pass to meet the needs of the downstream senior water users.

The miners developed a water rights system that accommodated the arid western landscape by allowing water to be diverted to lands miles from the nearest stream or river. Under this new system, monitoring and enforcement of water claims were more difficult. One advantage of the old riparian system was that it was relatively easy to determine who had a right to water because use was limited to those lands adjacent to the water source. Illegal or unauthorized use was more difficult to detect under the new prior appropriation system because water could be used virtually anywhere a ditch would reach. The system required constant monitoring of diversions to ensure that rights holders received the water to which they were entitled.

The Anderson-Hill model predicts that more resources are devoted to definition and enforcement when the probability of loss of an asset is high; monitoring of water diversions substantiates this prediction. Miners invented a device called a miners' inch box to regulate diversions. This was a simple device consisting of a board with one-inch-square holes cut in it that regulated the flow of water. The number of holes in the box reflected the quantity claimed by the rights holder: For example, a miner claiming three miners' inches of water would have three holes cut in the box. The device gave rise to the unit of measurement known as miners' inches, which is still used for many older water rights. While this method was easy to implement and was accurate, it was primarily used on heavily appropriated streams where water rights definition and enforcement were problematic due to the number of diversions.

These types of measures helped reduce conflict, but water disputes still occurred. To handle these disputes, mining camps organized committees and courts (Dunbar 1983, 61). This provided members of camps with a fast and low-cost way to resolve conflicts. The system worked well as long as disputes were among miners.

Water management and enforcement problems increased, however, with the introduction of irrigation as a competing water use with mining (Dunbar 1983, 61). Prior to irrigation, the water rights system used by miners was fairly informal. Miners based their claims on the capacity of diversion ditches to eliminate the need to formally record rights. This method made it difficult for new water users to identify water that was already claimed.

Mining benefited from the application of the prior appropriation doctrine because it predated most agricultural uses of water. As a result, miners held some of the best water rights because the doctrine's principle of first-in-time, first-in-right gives priority to the earliest uses of water. During times of drought, those with the oldest rights are entitled to divert water first until their needs are met. As the number of farmers in the West increased, so did their demand for water. As more water was diverted from streams for agriculture, miners argued that the additional water was already appropriated. The new demand by irrigators increased water scarcity and hence its value. In addition, the probability of loss of miners' rights had increased. As disputes grew, farmers and miners sought to formalize water rights.

Another example of increased effort to define and enforce water rights came in 1878. With competition for water between mining and farming particularly fierce, a group of water users, primarily farmers, held a statewide conference on recording, defining, and enforcing water rights. Most of the discussion during the conference centered on problems of determining priority dates and the allocation of water according to those priorities. As a result of the conference, a commission was established to record and monitor water rights. By law, water users were required to file water rights with the commission. Failure to comply resulted in a loss of priority (Dunbar 1983, 90–91).

The commission was also responsible for dividing the water of a stream among water users according to priority. In the event of shortages, they were authorized "to shut and fasten . . . the head-gates of any ditch . . . which . . . shall not be entitled to water by reason of priority of the rights of others below them on the same stream" (quoted in Dunbar 1983, 93). The approach developed in Colorado served as a model for other western states.

Water rights enforcement was also a central discussion at the 1878 conference. Unauthorized water use was a growing problem because most diversions were not monitored or gauged. It was easy and common for water users to divert more water than provided for by their rights. Elwood Mead, a prominent figure in the development of western water law, noticed this in 1886 while measuring and gauging diversions by irrigation ditches in the South Platte Valley. Noting wide discrepancies between the decreed and the actual amount flowing in ditches, Mead commented in a report, "that the results of the gauging and the decreed capacity seemed to have no connection with each other" (quoted in Dunbar 1983, 101).

Little has changed over the past century. Few diversions are gauged, and reports of illegal use are frequent. A common problem throughout the West is water spreading, the practice of using federally developed project water for unauthorized purposes or on unauthorized lands. The practice violates contracts that spell out exactly when, where, and how much water irrigators can use. According to a 1994 Department of Interior inspector general's report, taxpayers lost $46 million over nine years on twenty-four projects as a result of illegal water use (Koberstein 1994). The Bureau of Reclamation has taken steps to address the problem, but critics contend these efforts have done little to reduce water spreading.

States are under increasing pressure by environmentalists and water users to increase enforcement efforts. Both of these groups routinely point out that some irrigators divert far more water than allowed by their rights. For example, Water Watch of Oregon identified illegal water use in Klamath basin as a serious problem. Consistent with the predictions of the Anderson-Hill model, the Oregon Water Resources Department responded by ordering the installation of measuring weirs on all major diversions. Similarly, in 1996 the Idaho Department of Water Resources took formal enforcement action against more than fifteen landowners for using water without a valid right. Complaints by landowners that they were not getting all the water provided by their rights prompted an investigation that led to the charges.

Many incidences of illegal use stem from the fact that water rights are not clearly defined. For example, in the Idaho case, the illegal diverters claimed they were not aware that they were violating other water rights. Often these types of mishaps arise because of changes in land and farming practices that go unrecorded. These errors are typically realized when the water rights or the appurtenant land changes ownership.

OVERVIEW OF GEOGRAPHIC INFORMATION SYSTEMS

To avoid these problems, some states are employing GIS and other related technologies to enhance recording capabilities and to make water rights information and maps easily accessible.

The development of GIS in water rights definition is a logical extension in the evolution of water rights. Advances in personal computers have substantially lowered the cost of using the technology. The change in technology is also lowering the cost of water rights definition and enforcement. GIS is becoming a routine part of water rights management throughout the western states.

GIS combines spatial database management with statistical analysis and cartographic modeling. Users can layer many kinds of spatially referenced tabular information in a visual map format. GIS has emerged over the past decade as a powerful analysis and information management tool. Historically the use of GIS was limited to government agencies, universities, and large corporations that could afford the costly software, equipment, and specialized staff needed to operate a system. Affordable prices, increased computing capability of personal computers, and user-friendly software, however, have made GIS technology widely available to the private and public sectors. This allows better resource management where spatial variables are an important consideration. For example, GIS is used by air traffic controllers to guide planes through crowded airspace; by the United Parcel Service to track packages to their destination; and by local, state, and federal agencies to plan timber harvests (Chandler 1996). GIS is ideally suited as a water resource management tool because the vast majority of water resource data, including water rights, can be spatially referenced (Schoolmaster and Marr 1992; *Space News* 1995, 18).

Using GIS to Define Water Rights

In order to resolve water rights conflicts among owners and to allow for market trades, water rights must clearly specify the quantity of water that can be used and how and when it can be used. GIS is increasingly being used by state agencies to improve the way water rights are defined. The technology is enhancing access to information about ownership, use, and the quantity of water provided by the right. The following are examples from Idaho, Wyoming, and Oregon that illustrate how GIS is used to define water rights.

The Snake River Basin, Idaho

In 1982, the Idaho Supreme Court decision upheld a claim by the Idaho Power Company that its early 1900s water rights for 8,400 cubic feet per second of Snake River water was not subordinate to the water rights of upstream irrigators (*Idaho Power Company v. State of Idaho* 1982). The court's ruling severely overappropriated the Snake River basin, as existing water rights claims and pending claims exceeded the available water.

In response to the decision, the Idaho Power Company filed suit against 7,500 holders of pending water rights applications (Morse, Zarriello, and Kramber 1990, 365).[3] Facing millions of dollars in legal expenses and years of litigation, the state of Idaho and the Idaho Power Company negotiated a settlement. A condition of the agreement was that all water rights in the Snake River drainage would be adjudicated.

In ratifying the settlement, the Idaho legislature ordered the Idaho Department of Water Resources (IDWR) to provide the presiding court with all the technical information necessary for it to make a determination on each water right. This is an enormous task because the region includes more than 160,000 water rights within a 72,000-mile area (Morse, Zarriello, and Kramber 1990, 365). Spatial analysis tools such as GIS and remote sensing were ideally suited for this project.

IDWR developed a GIS program to aid the adjudication. GIS and remote sensing were used to estimate the irrigated acreage associated with each water right (Morse, Zarriello, and Kramber 1990, 366). These technologies facilitated a comparison of irrigated acreage with existing water rights and claims submitted as part of the adjudication. The quantity of water for water rights was established by the historic use of irrigated lands.

Under the prior appropriation doctrine, the measurement of both the amount stated in the right and the consumptive use of water by the rights holder are critical. This is especially true for water transfers. The amount of water that a rights holder can sell or transfer is limited by the amount he or she has historically used, regardless of the amount stated in a water right. For example, if a water user has a right to divert sixteen cubic feet per second, but has never taken more than eight cubic feet per second, the right is reduced to the amount historically applied to a beneficial use.

Initially IDWR used aerial photography to create digitized land classification layers that included irrigated lands. The land classification data were overlain with the legal property descriptions, allowing the department to estimate irrigated acreage for each sixteenth of a section (a forty-acre parcel).

This approach is limited by the course resolution of acreage estimates. For example, consider the situation in which there are four water rights on file, each claiming ten acres under irrigation. The course scale of these data cannot distinguish among these rights. Thus, field investigations are required to determine which part of the forty-acre parcel is covered by each right. The resolution quality was a serious problem for IDWR. Field investigation in the basin would be costly and possibly prohibitive because a large number of water rights are less than forty acres. As a remedy, the state devoted more resources to acquire satellite imagery with better resolution that allows individual water rights to be matched with a specific land area (*U.S. Water News* 1997).

GIS is a valuable management tool for IDWR because it allows the agency to provide the court with an estimate of irrigated acreage that is independent of claims from holders of water rights and the recorded water rights. The additional information supplied by GIS allows the court to process routine claims quickly, while identifying complicated and potential controversial claims that require field investigation. Several western states are using a similar process to adjudicate water rights and negotiate water rights settlements with American Indian tribes (Pearce 1999, 539).[4]

The Smiths Fork Irrigation District, Wyoming

The state of Wyoming experimented with GIS in 1998 to update and correct water rights information in the Smiths Fork Irrigation District (Wyoming Water Development Commission 1999). The demonstration project was designed to help the district map its water rights and correct discrepancies between the rights on file with the state. The district suspected that water use patterns had changed since the original rights were filed more than 100 years ago. To protect existing water rights from forfeiture due to lack of use, the district needed to correct discrepancies between the water rights records and actual use. Also, changes in landownership made it difficult to determine who held title to the land for assessment of water delivery charges.

The Smiths Fork Irrigation District entered into a cooperative agreement with the Wyoming Water Development Commission to map irrigated lands and create an overlay of existing water rights and permits. Using this information the district identified lands within its boundaries that were not irrigated and the extent of irrigation occurring outside its boundaries that were not covered by a water right. The district filed a petition with the state engineer to incorporate these lands within their boundaries. In addition, this project identified lands within the district boundary that were receiving water, but were not being assessed a delivery charge. GIS enabled district managers to calculate accurate delivery charges to members according to actual irrigated lands and ownership.

Oregon

The state of Oregon has an innovative approach to providing information on water rights that combines GIS technology with the instantaneous access capabilities of the Internet. This system allows users to easily produce maps of surface, groundwater, and instream flow rights as well as physical and natural features. The database, which is operated by the Oregon Water Resources Department, contains a broad array of information for individual water rights, including the area serviced by a right, the point at which the water is diverted, wells and pump sites for groundwater rights, and priority date information.[5] These can be used to select rights and generate maps based on location such as township, range, and section, or by time period according to priority dates.

The spatial database and mapping functions are popular with federal, state, and local government agencies, as well as the private sector. The Oregon Water Resources Department estimates that on average 350 maps are generated each month; approximately 70 percent of these are requested by people outside the agency. This indicates that a comprehensive GIS system can provide management assistance to a number of diverse groups. Public agencies are primarily using the online database as a research and management tool. For example, local watershed councils use the technology to identify the location of water rights on streams where cooperative water management plans are being drafted.

Within the Oregon Water Resources Department, field staff are using the mapping service to enhance management and enforcement efforts. The system allows them to plot and review water rights for an area before going into the field. They are able to build accurate and current maps for a region with multiple water rights in a matter of minutes, rather than having to construct one from individual water rights maps. The spatial database has enhanced enforcement efforts by providing field staff with maps that reflect current water rights information. Field staff rely on accurate water rights maps for investigating and detecting illegal or unauthorized water use. When conducting field investigations, the enforcement staff cross-reference the locations where they observe water use with water rights maps. Prior to the Web-based mapping service, staff used maps generated by the GIS division within the department. These provided up-to-date information, but there were usually lengthy delays before the maps made it to the field. As a result, most enforcement staff used old, often out-of-date maps on file with the water rights records.

The most frequent users from the private sector are real estate appraisers and agents. They use the online mapping system to describe water rights information in real estate valuations and transactions. The maps are commonly included in property descriptions that are sent to potential clients.

Digitizing the water rights records cost the state about $1.25 million; however, the water rights coverages are used in other capacities in addition to the Web-based mapping service, reducing information duplication and decreasing overall operating costs. In fact, the concept for the mapping service stemmed from an effort to make these data publicly available. The department spent approximately $10,000 in staff time to develop the online mapping service. Annual maintenance costs are minimal because the system is automated. Improvements in the system are ongoing and include the addition of more detailed water rights information, as well as increasingly sophisticated real-time, interactive mapping capabilities.[6]

Oregon's system is arguably the most advanced online spatial database and water rights mapping service, with Texas, Oklahoma, South Dakota, and Washington also offering Web-based mapping services. As more of these services come online, water rights holders will be able to take a more active role in water rights management and enforcement.

Using GIS to Enforce Water Rights

The Arizona Department of Water Resources was one of the first state agencies to use GIS and satellite imagery to detect illegal water uses. In the early 1980s, declining water levels forced the state to close several groundwater basins to additional appropriations. Enforcing the closure was particularly challenging because illegal pumping could occur anywhere in the basin. In the case of surface water, violations were relatively easier to locate because they were confined to a stream system. To improve enforcement, the state developed an approach that combines GIS and low-resolution infrared satellite imagery. Infrared imagery was used because areas with high vegetation growth could be identified. Water rights within the closed basins were mapped using GIS. Overlaying these coverages with the infrared satellite images allows the department to detect potential irrigation outside certified areas; the department then conducted field investigations for these areas. This process saved hundreds of personnel hours by eliminating the need for random field investigations (Pearce 1999, 538).

The use of satellite imagery with GIS to enforce water rights raised constitutional questions. In 1998, the Arizona Department of Water Resources fined a farmer $4,000 for illegally using water. The department employed satellite imagery with GIS water rights maps to show that the water user was irrigating lands not covered by his water rights. GIS was used to analyze the satellite images by overlaying water rights data with the images. The farmer readily admitted guilt, but when informed that satellite imagery was used in detection, he questioned whether such surveillance was a violation of his right to privacy (Kerber 1998).

Some legal scholars argue that expanding governmental use of satellite surveillance impinges upon their privacy rights secured by the Fourth Amendment (Kelly 1995; Kerber 1998).[7] Satellite surveillance may, in fact, constitute a warrantless and unreasonable search. Although courts have not squarely addressed this issue, relevant case law has developed in the context of aerial surveillance and photography (*Dow Chemical Co. v. United States* 1986).

In *Dow Chemical Co. v. United States*, the Supreme Court held that the Environmental Protection Agency (EPA) was not required to obtain a warrant before photographing a Dow Chemical facility. Dow Chemical argued that it had a reasonable expectation of privacy from airplanes equipped with technologically advanced cameras. On one hand, the court ruled against Dow Chemical on this issue, stressing that mere magnification of human vision does not give rise to a constitutional question. On the other hand, the court limited its holding by warning that the use of sophisticated technology not generally available to the public, such as satellites, may at some point require a warrant. Some reviewers argue that despite the court's warning, the current Fourth Amendment standards do not restrict warrantless satellite-enabled searches (Geer 1991, 46; Pearce 1999, 538). They contend that the degree of public access to a particular searching technology is the deciding factor (*California v. Ciraolo* 1986). Although satellite imagery was considered especially sophisticated in 1986 when the Dow Chemical opinion was written, today it is readily available to the public.[8]

Still others argue against the use of this sliding standard. The dissent in *Dow Chemical* (1986, 240), for example, pointed out that a standard that shifts with technological advance will quickly erode privacy rights. Other courts and legal commentators have agreed that this standard is a poor substitute for its predecessor—a standard that shifted in accordance with societal views rather than technological advances (Kelly 1995, 742). Finally, one might argue that merely requiring disinterested judicial permission for such searches does not overly burden state water enforcement officials.[9]

Regardless of the constitutional implications of the use of satellite imagery, much of the property involved in water rights allocation could be viewed without a warrant under the "open fields" and "plain view" doctrines (Kelly 1995, 753), under which many property owners may not have the required reasonable expectation of privacy. In any event, laws governing the use of satellite imagery will likely continue following aerial surveillance case law.

The controversy surrounding satellite surveillance may slow the use of GIS in water rights enforcement. Although Arizona is actively using the technology, other western states are approaching it with caution. The Wyoming Engineer's Office considered using GIS and satellite imagery for water rights enforcement, but the technology was never employed primarily because water rights holders and state policy makers raised objections over privacy concerns. Other state agencies have expressed similar worries. The issue of privacy must be addressed before GIS and satellite imagery become widely accepted as an effective water rights enforcement tool. Warrants are one solution to making the technology more socially and politically acceptable (Kelly 1995, 760).

GIS as a Tool for Water Marketing

The way in which water rights are assigned, enforced, and transferred affects the allocation of water. GIS has played an increasing role in both definition and enforcement, but has had limited application in water transfers. Yet GIS technology offers potential for the future of water trading.

Oregon's mapping service is a good example of how the technology can be used in an emerging market. Water users interested in acquiring additional rights use the mapping service to identify water rights near them that they could potentially purchase. With this information, a buyer is able to contact the current owners of water rights to negotiate a possible sale or transfer. The Oregon Water Resources Department is seeing an increasing number of people using the service for this purpose.[10]

The Oregon Water Trust also provides an example of how GIS can be used to plan water purchases.[11] The trust has limited financial resources to spend on water rights acquisitions, but has developed an acquisition strategy that incorporates GIS analysis to ensure that water rights purchases restore flows to protect fisheries habitat or improve water quality. They use a two-part strategy that involves selecting priority streams or stream reaches that provide substantial benefits to fish and then identifying water rights on those priority streams that effectively restore flows.

GIS is used to determine the ecological value of streams by evaluating current stream conditions such as fish distribution. Combining this analysis with additional information such as stream surveys, the trust is able to select streams that would benefit most from additional flows. Once a stream is selected and target flows are determined, GIS is used in the second stage to identify water rights that could be acquired to augment flows.[12] Targeted flows are usually greater than the amount of water provided by an individual right. Therefore multiple water rights need to be purchased to meet the flow objective. To accomplish this goal, the trust uses GIS to create maps that include information on flow levels at various points in a stream, the location of water rights diversions, the priority of those rights, and the current use of the rights. This allows the trust to compare junior and senior water rights to determine which are most cost effective in achieving desired flows.

Because the Oregon Water Trust uses a market approach to stream protection, their goal is to acquire the least-cost water rights that meet flow objectives. Often the organization attempts to restore flows in a specific section of a stream rather than throughout the entire system. In these situations, the most senior water rights on a stream are not always the most cost effective, especially given that senior rights typically cost more than junior rights (Landry and Mooney 1999). The trust can avoid paying a premium for rights with the highest seniority by acquiring rights with priority dates that provide the required level of protection at a lower cost. GIS aids in that selection process.

CONCLUSION

The level of enforcement and definition of water rights has changed considerably since rights were first established in the American West. Conflicts that arose over the increased scarcity of water drove many of the changes, and water rights are continuing to evolve through the assistance of GIS. The technology enhances the way information is gathered and recorded, and provides courts with a reliable and expedient method to verify water rights claims during adjudications. The technology is also improving the way water rights are monitored and enforced. State enforcement agents are able to access current water rights information, making it easy for them to pinpoint the location of water rights and monitor water use.

Improvements in the way water rights are defined and enforced will help only emerging water markets. Clearly defined, secure, and tradable property rights are fundamental to any well-functioning market. Because there are limited applications of GIS in water markets, it is not widely used to conduct transactions, yet the potential this technology offers is enormous. Through GIS and other associated technology such as satellite imagery, water rights holders can envision a day when they monitor and track the movement of water with the click of a button.

The same technology could be used to facilitate the exchange of water. Already, electronic water markets are taking shape over the Internet, where water is bought and sold. One of the advantages that electronic markets offer is that transactions will be instantaneous. These markets, however, will have to overcome regulatory obstacles that could offset any speed and efficiency gains offered by electronic transfers. At present, all water rights exchanges must undergo an administrative review. This process is often highly bureaucratic and can take weeks, if not years, to complete. The review is used to ensure that other water rights are not harmed by the proposed transaction. These delays will have to be reduced in order to realize the full benefits of electronic water markets. GIS could assist in cutting down the bureaucracy by helping to define areas where rights can be traded and the extent the rights can be moved without impairing other water rights. Water, unlike real estate, can be transported and moved to new locations. The technology provides a way for state agencies to offer advance approval for rights within a given market region. The preapproval would enable real-time electronic exchanges of water. The use of GIS is an example of innovations that will undoubtedly occur as markets emerge, the value of water increases, and more resources are devoted to not only defining and enforcing water rights but also improving the way rights are traded.

NOTES

1. In the summer of 1998, farmers in Klamath basin, Oregon, announced that, if necessary, they would use armed force to defend their water rights from being reduced to provide water for a federally protected fish (*Oregonian* 1998). U.S. Fish and Wildlife officials warned that such action would be prosecuted to the full extent of the law.

2. Self-regulated systems for common pool resources, such as water, function well when resource users are a homogeneous group. Conflicts tend to arise when new uses are introduced and narrowly defined norms can no longer be easily identified (Ostrom 1990, 88).

3. The company chose not to file suit on existing rights, determining that claims against pending applications would be more effective.

4. At present, no state has concluded a successful American Indian water rights claim using GIS or satellite imagery. Recent settlements have been under negotiations for years. GIS and satellite imagery will, however, play a greater role in future decrees (Pearce 1999, 539).

5. The Internet address for the water rights mapping database is: www.wrd.state.or.us/maps/index.shtml.

6. Bob Devylder, information technology director, Oregon Water Resources Department, Salem, telephone interview, October 5, 1999.

7. The Fourth Amendment of the U.S. Constitution affords individuals protection from unreasonable search and seizure of their "persons, houses, papers, and effects"; however, a judge may issue a search warrant upon a finding of probable cause. The premise behind obtaining a warrant on probable cause is that an impartial third-party judge balances between the individual's rights and the enforcement action of the state (Kelly 1995, 734).

8. In 1994 the United States lifted a ban on the sale of high-resolution satellite images. Since then, the industry has grown considerably with annual sales of $6 billion (Daratech 1998).

9. An American Bar Association task force is exploring the issue of using satellite images of private property without consent. The task force's recommendation is expected to influence how law enforcement and civil agencies use the images and at what point search warrants will be required.

10. Devylder interview, October 5, 1999.

11. The Oregon Water Trust is a nonprofit organization formed in 1993 that leases and purchases water rights for the purpose of increasing streamflows.

12. The trust uses other information such as field assessments, stream temperature models, and biological surveys for selecting streams and identifying targeted flow levels.

REFERENCES

Anderson, Terry L., and P. J. Hill. 1975. The Evolution of Property Rights: A Study of the American West. *Journal of Law & Economics* 18(1): 163–79.

Chandler, Montrese. 1996. Geographic Information System. In *Trends in Commercial Airspace*, ed. Keith Calhoun-Senghor. Washington, D.C.: U.S. Department of Commerce, Office of Air and Space Commercialization.

Daratech, Inc. 1998. Geographic Information Systems Markets and Opportunities. *Market Report*. Cambridge, Mass.

Dunbar, Robert G. 1983. *Forging New Rights in Western Water*. Lincoln: University of Nebraska Press.

Geer, Karen. 1991. The Constitutionality of Remote Sensing Satellite Surveillance in Warrantless Environmental Inspections. *Fordham Environmental Law Report* (fall): 43–56.

Kelly, Krysten C. 1995. Warrantless Satellite Surveillance: Will Our 4th Amendment Privacy Rights Be Lost in Space? *John Marshall Journal of Computer & Information Law* (summer): 729–62.

Kerber, Ross. 1998. When Is Satellite Photo an Unreasonable Search? *Wall Street Journal*, January 1, B1.

Koberstein, Paul. 1994. Water for the Taking: Some Irrigators Get Loose with the Law. *High Country News*, October 31.

Landry, Clay J., and Sian Mooney. 1999. Price Determinants: A Case Study of Oregon's Water Market. Working paper 99-10. Political Economy Research Center, Bozeman, Mont.

Morse, Anthony, Thomas J. Zarriello, and William J. Kramber. 1990. Using Remote Sensing and GIS Technology to Help Adjudicate Idaho Water Rights. *Photogrammetric Engineering and Remote Sensing* 56(3): 365–70.

Oregonian. 1998. Judge: Tribes, Fish Take Precedence in Water Dispute. June 10.

Ostrom, Elinor. 1990. *Governing the Commons: The Evolution of Institutions for Collective Action*. Cambridge, England: Cambridge University Press.

Pearce, Michael J. 1999. Analytic Geography in Western Water Management. *Natural Resources and Environment* (spring): 537–41.

Schoolmaster, Andrew F., and Paul G. Marr. 1992. Geographic Information Systems as a Tool in Water Use Date Management. *Water Resources Bulletin* 28(2): 331–36.
Space News. 1995. Oil Company Commissions Satellites to Spy for Water. September, 18–24.
U.S. Water News. 1997. GIS Data, Satellite Imagery to Become Part of Idaho Water Rights Adjudication Program. April.
Wyoming Water Development Commission. 1999. The Use of GIS Technology in the Smiths Fork Irrigation District. Report to the 1999 Wyoming Legislature, Cheyenne.

CASES CITED

California v. Ciraolo, 476 U.S. 207 (1986).
Dow Chemical Co. v. United States, 476 U.S. 227 (1986).
Idaho Power Company v. State of Idaho, Idaho S.Ct. 18 (1982).

3

Enforcing Property Rights in Western Water: Is It Better to Be Upstream with a Shovel or Downstream with a Model?

Robert B. Naeser and Mark Griffin Smith

The West's doctrine of prior appropriation is a property rights–based system (Anderson 1983). Historically, this system has been administered in the field by an extensive but unsophisticated network of simple measuring devices and the ditchriders, water commissioners, and state engineers who manage the system. Water is a property right, but also a fugitive resource. Western water rights are created by diverting or capturing the resource. A rights holder takes the water at the point of diversion, uses a portion of the diversion, and returns a portion to the stream as return flows. Given the simple flow-measuring devices under which western water law evolved, it was only possible to track one's right during the brief time it was in one's possession; the water belonged to the river before the diversion and after it was returned to the system.

New technologies fundamentally change this. Although it is still impossible to predict streamflow, technology provides opportunities to track property rights in the stream, not just at the point of diversion. Colorado water managers and users can now identify what water belongs to whom, where the water is within the system, and when it will arrive at different points along the river. In doing so, technology makes it possible to track a water right in its free-flowing, fugitive state. This fundamental change creates new opportunities for water conservation, water marketing, and instream flow protection.

This chapter focuses on two case studies in which technological innovations in monitoring and modeling are used to clarify property rights to water in Colorado. The two case studies concern intrabasin transfers by the city of Colorado Springs and adjudication of the Arkansas River between Kansas and Colorado. The case studies build in scale from a single but complex urban water supply system, to the allocation of water rights between states. The doctrine of prior appropriation creates the context for the two case studies.

THE PRIOR APPROPRIATION SYSTEM

The doctrine of prior appropriation has five basic tenets: seniority, beneficial use, diversion, forfeiture, and private property.[1] Seniority is the first principle, the one that gives the doctrine its name and the phrase most often associated with it, "first-in-time, first-in-right." Seniority is unrelated to location on a stream or river. Senior appropriators may shut off or "call out" junior appropriators in order of appropriation when they (i.e., the senior appropriators) are entitled to, but are not receiving, their allocated share of the river. The prior appropriation doctrine allows an appropriator to divert as much water as can be put to beneficial use. It requires the physical diversion of water from the stream for a right to be recognized and requires a water right holder to "use it or lose it." Water rights can be lost in most western states through abandonment or forfeiture. Hence, water rights are not associated with riparian land and, in fact, are not associated with land at all. They are real property rights that may be bought, sold, and transferred to new uses in new locations, subject to the provision that such a change in use will not injure other existing water rights.

Thus, for reallocation of water to occur via market transfer, there is a critical need to identify and prevent potential injury to existing appropriators (Howe, Schrmeier, and Shaw 1986). This in turn requires accurate quantification of historic consumptive use to ensure that other appropriators are not harmed. It is critical to note that a water right is not specified in terms of water used or consumed because not all water that is diverted is consumed. Some of the water that was originally diverted from the stream enters the soil, percolates down to the water table, and ultimately returns to the stream where it can be claimed by downstream appropriators. The importance of return flows in the appropriative system is significant. When water rights are transferred, the purchaser does not receive the entire divertable flow decreed to the original water right, only the consumptively used portion of that right.

The prior appropriation system also recognizes rights to water that is imported from another watershed by tunnels, pumps, and pipelines. This nonnative water is different from native water in that it can be used to exhaustion. Appropriators holding rights to nonnative water may use it over and over again until it is exhausted. In allowing nonnative water to be used to exhaustion, the law recognizes three facts. First, nonnative water is often developed at considerable effort and expense, so full benefit of the investment should go to the developer. Second, allowing other appropriators to benefit from nonnative return flows would result in their gain without compensation to those who made the investment. Third, allowing nonnative water to be used to exhaustion promotes water conservation.

Groundwater law, compared with surface water law, is more difficult to characterize. More than two groundwater doctrines are practiced in the United States, and their application is less than uniform across the country (Goldfarb 1988). Even within a single state, different laws will apply within different regions and even to different aquifers within the same region. Colorado's current groundwater law dates from 1973.[2] The law divides Colorado's subsurface waters into two primary categories, tributary and nontributary.[3] Tributary groundwater is hydrologically connected to a surface stream or other water body. Wells that cause depletions to a surface water flow are considered tributary wells. Tributary wells are required to have an approved augmentation plan for replacing material depletions to the surface stream, thereby protecting senior surface water rights. Augmentation plans generally require tributary well owners to buy existing senior surface water rights that are then no longer diverted, but are kept in the stream to make up for the depletions caused by pumping. Typically the amount of water required for augmentation is a small percentage of what is being withdrawn from the well because most of the well's yield comes from changes to aquifer storage.

Nontributary groundwater is defined as water that will, after 100 years of pumping, result in a less than 0.1 percent depletion to any surface stream. Nontributary wells are generally located at such a distance from a surface stream that interaction with the surface stream is extremely small or difficult to quantify. Nontributary wells are not administered by priority date.[4]

TWO WATER MANAGEMENT CASE STUDIES

Colorado sits in the heart of the American West in a hydrologic dilemma. Over the state's western slope, the high peaks of the Rocky Mountains block moist winter storms from the Pacific. The resulting precipitation falls over the upper Colorado River basin, where 60 percent of Colorado's water originates. It is hardly a problem that the Colorado basin should produce so much of the state's water except that 88 percent of the state's population resides outside the basin, primarily east of the Rocky Mountains along the Front Range and eastern plains.

The cities of the Front Range and the farms of the South Platte and Arkansas River basins are fundamentally dependent upon transmountain water, brought from the headwaters of rivers within the Colorado River basin. Since the late 1800s, seventeen ditches, channels, or tunnels have been constructed to convey water over or through the mountains such that approximately 27 percent of the water available to the eastern half of the state comes from the west slope (League of Women Voters of Colorado 1992). Although transmountain water has increased supplies and firmed up yields, Front Range cities have only recently begun to realize the full benefits available from transmountain water: It can be reused to exhaustion only if the water can be tracked. Water users, whether they are farms or cities, often hold rights for both native and nonnative water. They may use both types of water at the same time, but one type must be returned to the stream whereas the other may be used again. Further complicating the problem is the fact that the timing of water availability is as important as total supply. To water users in the arid West, water that comes when it is unneeded is more a torment than an asset. Thus, the temporal availability of nonnative flows also must be identified if they are to be sold or exchanged with other users.

Colorado water managers find themselves in the unenviable position of managing what little water is available in the face of growing demand, calls for environmental protection, and limited tax revenues. Water managers are required to identify new, renewable sources to meet projected water demands, as well as manage existing water rights within the structure of the prior appropriation system. The search is constrained by a general lack of unappropriated water and Colorado's compact obligations to downstream states.[5]

The solutions to the current water management issues lie, in part, within the prior appropriation system (Anderson and Snyder 1997). As previously mentioned, the property rights structure of the prior appropriation system allows for transfers of water rights, transbasin diversions, and the use to extinction of transbasin (nonnative) water. The average irrigated farmer in the intermountain West applies approximately two acre-feet of water to each irrigated acre in an average year (Economic Research Service 1993). Colorado farmers in the Arkansas River valley harvested approximately 272,500 irrigated acres in 1992 (U.S. Bureau of the Census 1992), enough water to potentially supply more than 1 million municipal customers for a year, hence the significant interest in agriculture to municipal water transfers. Moreover, municipalities are beginning to manage transbasin water such that it is used to extinction. This is not a trivial task because various sources of water must be completely tracked through the system to avoid harming other appropriators. Finally, water managers are finding additional water through careful management of the existing water rights system.

Two case studies presented in the following sections highlight the increased role technology plays in Colorado water management. The studies are organized in increasing order of geographic scope: local, regional, and interstate. As with water and the western economy, the two case studies have a symbiotic relationship; Colorado Springs is directly affected by the outcome of the *Kansas v. Colorado* (1995) case whereas water use in the lower Arkansas Valley depends upon management decisions by the city of Colorado Springs. The technology discussed may seem unsophisticated to one accustomed to life in the Internet age. But the transition from shovels, pickup trucks, and morning coffee at the local diner to personal computers, satellite gauging systems, and local area networks comes slowly to a system with its roots in the California gold rush.

Water Supply System for Colorado Springs

The municipal water supply system for Colorado Springs is a complex mix of native and nonnative water rights, exchange agreements, and groundwater.[6] In the 1980s, the city recognized that water management practices were not maximizing yield from the bundle of water sources in the city's water portfolio. Population growth coupled with the expense of developing new water supplies resulted in development of a new, technology-driven water management system, one that incorporates real-time satellite monitoring of diversions and river stages, real-time tracking of flows and storage within its delivery system, modeling of groundwater return flows, modeling of travel times for exchange releases downstream, and remote operational control of all major system components. The result is not only a higher yield to the city from its existing water rights but also the identification of flows that can be reliably exchanged or sold, potentially improving water use efficiency throughout the entire Arkansas River basin.

To manage this complex system, Colorado Springs has developed a state-of-the-art computer-based water supply management system. The system consists of three principal components. The first is a satellite monitoring system called Sutron that has been operational since the early 1990s. Sutron monitors flows at all of the city's diversion points and all major stream gauges within the watersheds of its supply system. It also monitors the water surface elevation of all storage reservoirs. Measurements are made at fifteen-minute intervals and integrated to daily values. Sutron provides the city with nearly real-time data on inflows and storage in the system and triggers out of compliance warnings on minimum streamflows.

The second component is the supervisory control and data acquisition (SCADA) system. SCADA allows the operations control center in Colorado Springs to remotely operate all major components of the system through remote telemetering units (RTUs). SCADA not only allows operators to monitor and control raw water supply, finished water delivery, and wastewater flows from a single location but also open and shut valves, bring hydropower units online, and control raw or finished water storage. SCADA is not unique to the Colorado Springs system; many municipalities rely upon SCADA systems to monitor storage reservoir levels.

The third component of the system is the water information system (WIS), which integrates data generated by Sutron and SCADA to track diversions, raw water storage, finished water production and storage, wastewater inflows, nonpotable water production and storage, and effluent return flows. WIS tracks the quantity of water entering the system and the quantity stored, used, and then returned to the stream on a daily basis. Moreover, WIS tracks the origin of all water flowing through the system, monitoring whether it is native or nonnative. This task is complicated by the fact that water from a single reservoir might belong to one of seven different accounts that can be tracked to the hundredth acre-foot.[7] More than a highly sophisticated inventory system for water supply accounting, the WIS is also a decision support system.[8] Municipal irrigation (i.e., landscape watering) is a significant component of summer water use. Although it is possible to measure outdoor water use as the difference between finished water production and wastewater inflows, such a rough measure of water use does not account for the difference between use and consumption. Not all water that is applied to lawns and gardens is consumed through evaporation and evapotranspiration. Often, more than half of what is applied will not be consumed, but will return to the system as return flows. The question is when. The WIS tracks return flows through a groundwater return flow model integrated into WIS. Thus, WIS tracks not only water that has been in the conveyance system for less than forty-eight hours but also water that will return to the stream days, weeks, even months later as groundwater return flows. WIS allows detailed quantification of property rights to native and nonnative water so that the city can maximize economic and physical benefits from its limited supply.

Benefits Realized from a Technology-Driven Water Management System

The most recent legal innovation in maximizing the value of nonnative water rights is an exchange agreement that developed out of the city's ability to track water from different sources throughout its transit through the system. An exchange agreement is a physical trade of water between two appropriators. In an exchange agreement, one party with native direct flow rights agrees to trade native water for nonnative water released by the other party. An exchange agreement requires the party releasing the nonnative water to match the availability of water from the native source in both time and quantity.

For example, Colorado Springs can exchange transmountain water for senior native water with a number of mutual ditch companies that hold direct flow rights in the lower Arkansas Valley. Historically, these canal companies would take all of their direct flow water rights from native water originating in the Arkansas River basin. With an exchange agreement, Colorado Springs diverts native flow near the river's headwaters and pumps it through a conduit to the city. If Colorado Springs diverted this water without replacing it, the canal companies would receive less than their entitlement under the doctrine of prior appropriation, but Colorado Springs replaces the water by releasing an equivalent quantity of transmountain water. The exchanged water reaches downstream headgates at the same time that the native flows would have been available had they remained in the Arkansas River.

Such an exchange allows the city to optimize its water rights in several ways. First, the law allows that nonnative water may be used to exhaustion. Some of the water the city releases to fulfill the exchange agreement is nonnative water, which is now nonpotable sewage effluent. The exchange agreement allows this water to be used twice. In addition, some of the water counted in fulfilling the agreement is groundwater return flow from landscape use within the city. Since these return flows would be almost impossible to recapture, the exchange agreement allows for these rights to be fully utilized by exchanging them to a downstream user. Second, the water released to meet the exchange is nonpotable. Nonpotable water is expensive to treat by a municipality, but perfectly adequate for irrigation. Third, the exchange agreement allows the city to divert native flows from the Arkansas near its headwaters where water quality is high, pumping costs are lower, and the delivery system is already in place. Exchange agreements give the city flexibility to trade nonpotable water for high-quality native flows.

Exchange agreements are legal documents that must be approved by the water court. Implementing an exchange agreement requires the party applying for the exchange to show no injury to any vested water rights. One of reasons that exchanges can work on the Arkansas River is that most of the senior water rights on the river are in the farming communities in the lower valley. Therefore, flows in the upper reaches of the river can be depleted without injuring water rights between the pumping station near the headwaters and irrigators 100 miles downstream.

In addition to such exchange agreements, Colorado Springs sells sewage effluent and irrigation return flows emanating from its transmountain water rights to downstream irrigators. Demand for these flows comes from two markets. First, irrigators holding Arkansas River direct flow rights need to firm up their water supply when natural flows are, or are forecasted to be, inadequate. Prices for Colorado Springs's return flows increase over the irrigation season. Lately the market price has started at $6 per acre-foot in April and May, rising to $8 per acre-foot in June and July, and then to $12 per acre-foot in August through the end of the irrigation season. Prices rise over the irrigation season because Colorado Springs loses the opportunity to refill reservoirs once the spring snowmelt ends.

The second market is for augmentation water needed by the Arkansas Valley's two groundwater user associations, which, under the recently settled Arkansas River Compact lawsuit, must restore flows across the state line into

Kansas. In Colorado Springs, return flows are sold annually. The city reserves the right to terminate sales when it becomes cost effective to treat and reuse this water. From 1997 to 1999, the city used, sold, or exchanged an average of 74 percent of its sewered return flows.

From 1997 to 1999, Colorado Springs sold an average of 25,000 acre-feet per year to nine different buyers, enough water to annually serve a city of 100,000. Buyers include four mutual ditch companies acquiring supplemental irrigation supplies and the two groundwater user associations purchasing water for well augmentation.[9] In addition, the state of Colorado has annually purchased water from Colorado Springs for wildlife and instream flow purposes. The Division of Wildlife buys water to maintain waterfowl habitat in the lower Arkansas Valley, and the Division of Parks and Outdoor Recreation must acquire water to maintain summer whitewater recreation on the upper Arkansas.[10]

Technology has been the key to full utilization of transmountain water. The Colorado Springs water management system integrates real-time satellite monitoring of diversions and river stages, real-time tracking of flows and storage within its delivery system, modeling of groundwater return flows, modeling of travel times for exchange releases downstream, and remote control of all major system components. The result is not only a higher yield to the city from its water rights, but also the identification of flows that can be reliably exchanged or sold to downstream users.

Interstate Water Management between Kansas and Colorado

The recent Arkansas River Compact dispute between Kansas and Colorado serves to illustrate the role technology now plays in litigation, as well as monitoring and enforcing property rights after a case has been decided.[11] In this example, technology was instrumental to defining both Kansas's and Colorado's property rights and ensuring that Colorado remains in compliance. Moreover, the decision created conditions in which a market could develop.

History of Arkansas River Disputes

Apportionment of Arkansas River flows was the focus of conflict between Kansas and Colorado four times in the twentieth century: *Kansas v. Colorado* (1902); *Kansas v. Colorado* (1907); *Colorado v. Kansas* (1943); and *Kansas v. Colorado* (1995). The driving force in each of the four cases was Kansas's claim that Colorado irrigators were using more than their equitable share of Arkansas River flows. Kansas failed to prove, in the earlier cases, that Colorado had materially decreased Arkansas River flows and that the alleged decreases worked to the detriment of Kansas's interests.[12]

To help resolve conflicts, the Arkansas River Compact was ratified by the legislatures of Kansas and Colorado and the U.S. Congress in 1949.[13] But even this was not enough because the compact did not quantify the water rights for each state (Bennett and Naeser 1997).

As a result, Kansas filed suit against Colorado in 1985 claiming that increases in postcompact alluvial (i.e., tributary) groundwater pumping by southeastern Colorado irrigators caused a significant decline in usable state line flows. The Supreme Court–appointed special master[14] decided in favor of Kansas on the issue of postcompact alluvial well development in Colorado and required southeastern Colorado irrigators to cease pumping or find replacement water for approximately 2,000 tributary wells that are junior to Colorado's obligation to Kansas (Robbins 1996). Restitution will be determined by the special master in the upcoming damage phase.

The Role of Technology in Kansas v. Colorado *(1995)*

Kansas and Colorado developed Arkansas River flow simulation models to support their positions in the case.[15] The litigation models were instrumental in proving that postcompact groundwater development in Colorado did result in diminished usable flows to Kansas irrigators. Ultimately, the Hydrologic-Institutional Model (HIM) developed by Kansas was used to simulate surface water flows in a geographic region limited to the main stem of the Arkansas River from Pueblo, Colorado, to the Colorado/Kansas state line, including tributary groundwater aquifers. The model estimates a water budget[16] based on water consumptively used versus return flows via surface water and groundwater paths. Groundwater return flows are lagged to reflect the fact that it can take days, weeks, months, or even years before groundwater impacts are realized in the surface stream. The lag period varies across the modeled region depending on proximity of the groundwater diversion to the surface stream and local hydrogeological factors. The surface and groundwater components are indirectly linked within the HIM. These modules generally operate independently of one another until groundwater flows connect with the surface water system. Based on this model, Colorado must manage its water so as not to materially deplete water flowing to Kansas.

The *Kansas v. Colorado* (1995) decision has fundamentally changed water management in the Arkansas River basin. Colorado now has a defined management target—the river entering Kansas must be in the same condition as existed prior to 1950. Thus, all postcompact groundwater development is in violation of the compact; however, significant personal and regional economies are based on this ill-gotten water. The state is attempting to proactively manage the Arkansas River, realizing that established demands currently exceed its legal entitlement. Technology is playing a pivotal role in this process.

In response to the verdict, Colorado developed a model to track depletions to state line flows as the difference between groundwater pumped and surface water augmentation. Irrigators know how much surface water they will be charged for groundwater depletions. Thus, these appropriators have the necessary information to either change irrigation practices to reduce presumptive depletions or go into the market to purchase replacement water. Although no evidence exists that irrigation practices have changed, these appropriators are turning to the market for replacement water, which can take the form of senior surface water rights, mutual ditch company shares, or nonnative return flows. Senior surface water rights are the most expensive, particularly in light of the low cost of return flows from the Colorado Springs system. But the time may come when Colorado Springs ceases to sell return flows and the market for senior rights and mutual ditch shares will dictate whether these irrigators make the decision to bid against municipal water interests, go dry, or negotiate interruptible supply agreements.

Since 1996, their first year operating under the special master's decision, the Arkansas Groundwater Users Association has purchased an average of 8,000 acre-feet per year for well augmentation at an average of $8 per acre-foot. Colorado Springs is but one source; the Southeastern Colorado Water Conservancy District and Pueblo Water Works sell in this market as well. Contract terms vary widely across the sellers. Colorado Springs sells its return flows that come down the river as they are released by the city. Thus the city offers only a take or pay contract for its return flows. Because the conservancy district has storage, their water can be called for on an as-needed basis. In addition to these annual purchases, the association has also purchased one-third of the Excelsior Ditch as a permanent source of augmentation water.[17]

Water management changes arising from the 1995 case are primarily limited to groundwater use. Economic efficiency would see water move from lower- to higher-value agricultural uses.[18] At the margin, however, agriculture is no longer the highest-value water use in southeastern Colorado. At the beginning of the twenty-first century, irrigators are one of many groups interested in river flows. Recreationists on the upper reaches are demanding flows to maintain aquatic habitat and enhance the recreation experience, and cities along Colorado's Front Range have paid up to $8,000 per acre-foot for permanent transfer of Arkansas River water rights (Naeser and Smith 1995). Therefore, if values to other uses are considered, the verdict and the resulting administrative changes may be suboptimal if they reduce the amount of water available for higher-value uses.

OTHER TECHNOLOGICAL PROSPECTS FOR WATER RIGHTS SPECIFICATION

Colorado has recently developed another computer-based system for managing and modeling water use in the Colorado River basin, the Colorado River Decision Support System (CRDSS). The CRDSS, although new and generally unproven, has the potential to be used to identify opportunities for water rights transfers and exchanges in a basin that plays a critical role in both Colorado's water use and use by downstream states, including California and Nevada.

In the late 1980s, Colorado's water resources were coming under pressure from federal programs, other states, and increased demands within the state.[19] Kansas was alleging compact violations by Colorado in the Arkansas River basin. Analysis of Colorado River Compact obligations was becoming increasingly difficult using traditional approaches under court-imposed timelines. Recovery of endangered aquatic species within the Colorado River basin, and other basins, was also becoming a significant issue.

Colorado water managers recognized that solutions to these issues required cooperation among state agencies, water providers, and water users. The state determined that a tool was needed to provide stakeholders with better means of organizing, accessing, and analyzing available data and alternative management strategies. The state decided that a comprehensive, real-time water management system with predictive capability could greatly assist water managers and users with the complex issues facing the basin.

The primary goal of the CRDSS is to provide a tool with which decision makers can access credible information and make informed decisions concerning management of water in the Colorado River basin. To this end, the CRDSS integrates water resource data and models into a functional system that provides the capability to develop informed management decisions based on the best available data. The secondary goal of the CRDSS is managing limited water resources without additional employees, a necessary condition under the current, fiscally conservative administration.

These two goals are met by incorporating a relational database (HydroBase), four simulation models (Colorado River Basin Simulation Model, the Water Resource Planning Model, the Consumptive Use Model, and the Water Rights Administration Tool), and a geographic information system (GIS) to meet the system's stated goals. The CRDSS was designed to allow integration of more modules as additional needs are identified. The CRDSS shell and modules were designed to facilitate development of decision support systems in the state's three other river basins using a common structure. The CRDSS, including all databases and simulation models, is accessible via the Internet (http://cdss.state.co.us) to all interested parties with access to a personal computer and Internet connection.

The CRDSS provides benefits to the Colorado Division of Water Resources (DWR) and other water users in the basin. Benefits accrue to DWR through increased administrative efficiency. The CRDSS enhances the agency's ability to fairly and efficiently administer the prior appropriation system. The HydroBase component of the CRDSS includes historic data on streamflow, reservoir levels, water rights, and climate. Historic information is augmented with real-time data pertaining to streamflow, reservoir levels, climate, and snowpack data. These data have always been available to river managers but previously required telephone calls to numerous agencies (e.g., the U.S. Bureau of Reclamation, the state Division of Water Resources, the U.S. Natural Resources Conservation Service) and, perhaps, a drive to look at the river. The CRDSS maximizes administrative efficiency by providing a single source for real-time data and access to analytical tools.

Other water users receive benefits from increased administrative transparency and efficiency, as well as free data access. One of the basic purposes of the CRDSS is information sharing through an open system. Water users have access, through the CRDSS, to the same real-time information used by water commissioners to administer water rights. Increasing the transparency of the administrative process has the potential to alleviate the mistrust that traditionally pervaded the prior appropriation system.[20] If an appropriator is confused by an administrative decision, the data behind the decision are available for review through the CRDSS. Administration decisions, using the CRDSS, are based on the best available data and eliminate some of the guesswork that may have characterized water management in the past. Transparency can cut both ways in that water managers and users in other states also have free access to the CRDSS. For example, lower basin states can access the CRDSS to learn how much water is being consumed in Colorado and use the models to understand how Colorado DWR administers the river. The benefits of an open system, however, are perceived by Colorado to outweigh the costs because free access to the CRDSS may curtail future interstate disputes by improving water management efficiency within the state.

In addition to the administrative benefits, the CRDSS has the potential to increase economic efficiency. The prior appropriation system provides for a limited transfer of rights, but transfers typically involve high transaction costs (Getches 1990). For example, a senior appropriator may use a senior water right to irrigate low-value crops. A municipal or industrial interest may be interested in purchasing this water right for a higher value use. The water rights transfer, while theoretically simple, is often complicated by legal restrictions and transaction costs. Transaction costs arise because the transferee is required to demonstrate that other appropriators will not be harmed, and other appropriators may respond with legal challenges. Further, the nature of water rights makes it difficult to prove how much water can be transferred (i.e., consumptive use).

The combination of the Consumptive Use Model and Water Resource Planning Model provides agencies and water rights holders with tools for estimating potential effects from water rights transfers. The Consumptive Use Model affords stakeholders the opportunity to estimate the historic consumptive use under a particular water right. The change in diversion can be fed through the Water Resource Planning Model to estimate impacts on other appropriators. Although actual consumptive use may vary from the amount predicted by the model because of differences in annual cropping patterns, irrigation methods, and variations in local environmental conditions, modules within the CRDSS provide interested parties with a standard, costless (except for time) baseline impact assessment. The CRDSS provides a double check and a standard against which other calculations can be compared.

Water rights administration using the CRDSS will potentially result in securing the value of lower-priority rights. These lower-priority rights receive water only after demands by senior appropriators are satisfied. As water rights administration becomes more efficient using the CRDSS, DWR hopes to maximize allocation efficiency and conservation, thereby identifying water available for lower-priority users.

The CRDSS enables Colorado water users to maximize benefits available from water rights by providing improved access to data and increasing the transparency of the water rights administration process. Further, the CRDSS lets decision makers simulate potential management decisions and administer the prior appropriation system using current information. Ultimately the CRDSS has as much potential for promoting water markets as for improving state administration of water rights. The CRDSS puts information into the hands of potential buyers and sellers as well as enabling them to model the impact of sales on the river. By reducing information costs and identifying possible third-party injury, the CRDSS lowers barriers to market activity.

CONCLUSION

The two case studies discussed here demonstrate how technology allows appropriators to clearly define and protect their property rights in water. As currently used, these technologies ensure that appropriators receive their full, legal entitlement by reducing leakage in a system that historically resulted in suboptimal use and administration of existing water rights. In the Colorado Springs example, the WIS permits tracking and reuse of return flows from transbasin water diversions. The use of technology in *Kansas v. Colorado* (1995) resulted in quantification of the two states' respective property rights in Arkansas River flows and Colorado's future compliance with the Arkansas River Compact.

These case studies further demonstrate the role of technology in creating new markets. Failure for markets to form is commonly attributed to one of four inefficiencies: asymmetric information, externalities, public goods, or transaction costs (Rosen 1995). Asymmetric information occurs when one party in a transaction holds information not available to the other. Externalities exist when individuals make decisions that affect the welfare of others in a way that is not captured by established markets. Public goods are those that are nonrival in consumption; that is, one person's consumption of the resource has no bearing on another's use. Transaction costs arise when the cost of completing the exchange (e.g., collecting information, identifying willing buyers/sellers, legal costs, etc.) exceeds the gains from trade.

Anderson and Leal (1991) suggest that markets have not developed for some natural resources because the costs of monitoring and measuring the resource are high. The key to establishing markets comes from the establishment of well-specified, defensible, and transferable property rights. Theoretically, water rights under the prior appropriation system are well specified, defensible, and conditionally transferable. Anderson and Leal postulate that the historic lack of broad-scale water markets occurred because the nature of the resource makes monitoring and enforcing water use difficult, and the prior appropriation system imposes significant transaction costs on potential exchanges by requiring that all transfers be adjudicated in water court.

Technology also opens opportunities for addressing instream flow protection, one of the West's most contentious water management issues. A water right has traditionally been created by capturing water through diversion and putting it to a beneficial use. Instream uses such as habitat protection, recreation, and aesthetic values are not, however, included in the definition of beneficial use. Nevertheless, demand for protecting wildlife and its habitat, scenic areas, and environmental quality is increasing. Maintaining streamflow is often an integral part of sustaining environmental quality. Throughout the West, people are discovering innovative means of saving streams by maintaining instream flows (Landry 1998). The technological approaches presented in the case studies make it possible for instream flow rights to be defined and tracked. The Colorado Water Conservation Board uses the CRDSS to monitor instream flow rights in the Colorado River basin.[21] The next step is to allow anyone, from fishing clubs to the Sierra Club, to hold an instream flow right itself or, rather than acquire the right, purchase an easement on a downstream right to maintain that historic point of diversion (Anderson and Johnson 1986).

Technology removes one of the traditional constraints to trade in water rights, namely that each party incurred significant expenses attempting to monitor and measure the activities of the other (Anderson and Leal 1991). As demonstrated in the two case studies, technology can be used to clarify and monitor property rights to return flows in the Colorado Springs system, and it can be used to specify allocations between states, as in the dispute between Kansas and Colorado. If property rights can be efficiently monitored and defended once they are clearly defined, water managers and other stakeholders are positioned to enter into mutually beneficial arrangements: instream flow rights, lease agreements between cities and agricultural interests, exchange agreements, or full-scale water markets. Technological advances in water management make it increasingly better to be downstream with a model than upstream with a shovel.

NOTES

The authors gratefully acknowledge Ray Bennett, Jeanette Bryan, Scott Howell, David Robbins, Jim Slattery, Dale Straw, and Steve Witte for providing information, data, and perspective. David Preston provided invaluable production assistance.

1. The doctrine of prior appropriation as described here is known as the Colorado Doctrine and is used in Colorado, New Mexico, Wyoming, Montana, Idaho, Nevada, and Arizona. The relatively wetter coastal states, California, Oregon, and Washington follow combined elements of both the riparian and appropriative systems and are known as California Doctrine states. For differences in the two doctrines, see Goldfarb (1988). A good reference on the doctrine of prior appropriation is Rice and White (1987).

2. 37-90-91-91-102(7) C.R.S. (1973).

3. Colorado groundwater law recognizes a third groundwater category, not nontributary groundwater (Grantham 1989). This category includes aquifer formations that are deep or remote from surface streams and are administered separately from other aquifers in the state, such as Colorado's Denver basin. Not nontributary groundwater was excluded from this discussion of groundwater law because it does not pertain to the case studies presented.

4. All categories of groundwater wells are administered under a permit system that specifies the location of the well, the time of year it can be used, its rate of discharge, and whether it is tributary, nontributary, or not nontributary.

5. Unappropriated water still exists in Colorado; however, these sources are geographically isolated from demand.

6. Technical aspects of the water supply and water management system for Colorado Springs were provided by Scott Howell, Colorado Springs Utilities Water Department, in an interview with the author, July 16, 1999.

7. Water stored at Twin Lakes Reservoir, for example, includes the following accounts: Twin Lakes Native, Twin Lakes Exchange, Blue Exchange, Twin Transmountain, Colorado Canal, Homestake Exchange, and Colorado Canal Exchange. These rights represent not only native and nonnative water but various appropriation dates as well.

8. A decision support system integrates one or more computer-based models with their corresponding databases to allow a user, sometimes a nonspecialist, to evaluate alternative scenarios.

9. Scott Howell, Water Rights Administrator, Colorado Springs Utilities, e-mail, March 6, 2000.

10. The Arkansas Headwaters Recreation Area has an agreement with the Bureau of Reclamation and the Southeastern Colorado Water Conservancy District whereby the bureau maintains adequate flows in the upper Arkansas River to allow whitewater rafting through August 15 (see Naeser and Smith 1995). This modification to the bureau's reservoir operating plan results in increased evaporation losses when water is moved earlier in the season from the bureau's high-elevation reservoir at Twin Lakes to its lower-elevation reservoir at Pueblo. Water purchased from Colorado Springs is used to make up for these evaporation losses.

11. This case study relies heavily on material drawn from Naeser and Bennett (1998) and Naeser (1997).

12. For a more detailed discussion of these early cases, see Naeser (1997) and Wagner (1984).

13. Arkansas River Compact, ch. 155, 63 Stat. 145 (1949); 1949 Colo. Sess. Laws 485, 1 (codified at C.R.S. 37-69-101[1973]); 1949 Kan. Sess. Laws 829 (codified at Kan. Stat. Ann. 82a-520).

14. A special master is a specialist in water law appointed by the Supreme Court to hear an interstate water case. The special master is usually a highly regarded attorney or law professor.

15. Information on the Kansas Hydrologic-Institutional Model and Colorado's simplified groundwater accounting model was provided to Robert Naeser in interviews with Steve Witte, division engineer, September 13, 1999, and Dale Straw, assistant division engineer, Colorado DWR Division 2, by telephone, and Jim Slattery, Helton & Williamson, by telephone, October 6, 1999.

16. A water budget, or hydrologic balance, uses a mass balance approach to quantify the change in storage within a basin as the difference between inflows (e.g., precipitation, transbasin water, and groundwater outflows to gaining streams) and outflows (e.g., consumptive use, groundwater outflows from losing streams, and surface water flowing out of the basin).

17. Jeanette Bryan, manager, Arkansas Groundwater Users Association, telephone interview, March 3, 2000.

18. Preliminary analysis of average irrigation water values presented by Naeser and Bennett (1998) was inconclusive with regard to which region has higher water values. Average water values calculated for a bundle of three low-value irrigated feed crops (corn for grain, sorghum, and wheat) using the short-run crop budget approach were $45 in southeastern Colorado and $62 in southwestern Kansas in 1995 dollars.

19. Information on the Colorado Decision Support System was obtained from Ray Bennett, special projects manager, Colorado State Department of Water Resources, Denver, personal interview, June 3, 1999, and from the CRDSS Website: http://cdss.state.co.us.

20. This point was emphasized by David Robbins, a noted water lawyer and Colorado's lead council in *Kansas v. Colorado* (1995), during a personal interview in Denver, June 3, 1999.

21. In Colorado, only the Colorado Water Conservation Board can hold instream flow rights.

REFERENCES

Anderson, Terry, ed. 1983. *Water Rights: Scarce Resource Allocation, Bureaucracy, and the Environment.* San Francisco: Pacific Research Institute for Public Policy.

Anderson, Terry L., and Ronald N. Johnson. 1986. The Problem of Instream Flows. *Economic Inquiry* 24: 535–54.

Anderson, Terry L., and Donald R. Leal. 1991. *Free Market Environmentalism.* San Francisco: Pacific Research Institute for Public Policy and Westview Press.

Anderson, Terry L., and Pamela Snyder. 1997. *Water Markets: Priming the Invisible Pump.* Washington, D.C.: Cato Institute.

Bennett, Lynne. L., and Robert B. Naeser. 1997. The Arkansas River Compact: Integrated Watershed Management as a Means of Conflict Prevention. Paper presented at American Water Resources Association/Universities Council on Water Resources Symposium, "Water Resources Education, Training, and Practice: Opportunities for the Next Century," Keystone, Colo., June 29–July 3.

Economic Research Service. 1993. *Agricultural Resources: Cropland, Water, and Conservation Situation and Outlook Report*, AW-30. Washington, D.C.: U.S. Department of Agriculture, May.

Getches, David. H. 1990. *Water Law in a Nutshell*. St. Paul, Minn.: West Publishing.

Goldfarb, William. 1988. *Water Law*, 2nd ed. Boston: Butterworth.

Grantham, Joseph. 1989. *Synopsis of Colorado Water Law*. Denver: Colorado Division of Water Resources.

Howe, Charles W., Dennis R. Schrmeier, and W. Douglas Shaw Jr. 1986. Innovative Approaches to Water Allocation: The Potential for Water Markets. *Water Resources Research* 22(4): 439–48.

Landry, Clay J. 1998. *Saving Our Streams through Water Markets*. Bozeman, Mont.: Political Economy Research Center.

League of Women Voters of Colorado. 1992. *Colorado Water*. Denver.

Naeser, Robert B. 1997. Things to Do Before the Well Runs Dry: Estimation of Irrigation Water Values Along the Middle Arkansas River Valley. Master's thesis, Yale University.

Naeser, Robert B., and Lynne L. Bennett. 1998. The Cost of Noncompliance: The Economic Value of Water in the Middle Arkansas River Valley. *Natural Resources Journal* 38(3): 445–63.

Naeser, Robert B., and Mark. G. Smith. 1995. Playing with Borrowed Water: Conflicts over Instream Flows on the Arkansas River. *Natural Resources Journal* 35(1): 93–110.

Rice, Leonard, and Michael D. White. 1987. *Engineering Aspects of Water Law*. New York: John Wiley.

Robbins, David. 1996. What Did the Court Say? Address to the annual meeting of the Arkansas River Basin Forum, "A River of Dreams and Realities," Pueblo, Colo., January.

Rosen, Harvey. S. 1995. *Public Finance*. Chicago: Richard D. Irwin.

U.S. Bureau of the Census. 1992. *1992 Census of Agriculture*. Washington, D.C.

Wagner, Mark J. 1984. The Parting of the Waters: The Dispute between Colorado and Kansas over the Arkansas River. *Washburn Law Review* 24(4): 99–120.

CASES CITED

Colorado v. Kansas, 320 U.S. 383 (1943).

Kansas v. Colorado, 185 U.S. 125 (1902).

Kansas v. Colorado, 06 U.S. 46 (1907).

Kansas v. Colorado, 514 U.S. 673 (1995).

4

Using Geographic Information System Mapping and Education for Watershed Protection through Better-Defined Property Rights

Barrett P. Walker

The problem of open access to a common resource was described by Garrett Hardin (1968) in his classic article, "The Tragedy of the Commons." The often-told story from the article is the tale of the pasture open to all. According to Hardin, such an arrangement may work well for centuries because tribal wars, poaching, and disease may keep the numbers of cattle and herdsmen from reaching their natural limits. Eventually a desired period of stability arrives and each herdsman seeks to maximize the size of his herd. The individual herder's decision is a simple one. If he grazes an additional cow on the village green, he benefits from the sale of the animal. If he defers grazing one more animal, the pasture is less likely to suffer the effects of overgrazing, but the negative effect of overuse is spread among all the herdsmen. The most rational decision for the herder is to graze one more animal on the green, but all the herdsmen continue increasing the size of their herds until the green is ruined. In Hardin's words, "Each man is locked into a system that compels him to increase his herd without limit—in a world that is limited. . . . Freedom in a commons brings ruin to all" (1244).

The tragedy of the commons has been averted for farmland through the institution of private property. Secure and tradable property rights provide today's cattle ranchers with an incentive to balance current income derived from grazing a large herd against the future value of their land. If they overgraze, their income may rise for a few years, but the value of their land will fall.

Air, water, and biodiversity are not as well protected from the tragedy of the commons under current law. Our particular concept of property does not extend to the atmosphere, water that flows across the land, or species that have territories different than our own. It is much harder to establish property rights for air, water, and species that cannot be fenced.

Public bureaucracies have traditionally managed urban water resources through a fragmented, compartmentalized approach that treats storm runoff as a waste product. But the hydrological cycle is too complex to respond to such rigid, compartmentalized management. This chapter describes two projects that use geographic information system (GIS) technology to create property rights for water resources, beginning with neighborhood streams that supply rivers, lakes, and estuaries—components of the hydrologic cycle essential to our well-being. It shows how a free market approach to environmental protection can be implemented through the use of information technology that integrates economic and ecosystem management. In the past, complicated patterns of property ownership and resource distribution within urban drainage basins made practical application of free market environmentalism hopelessly complex. Despite its inefficiencies, command-and-control regulation was often the only practical approach for controlling pollution, yet poor urban water quality is fundamentally a result of habitat loss and altered hydrology caused by stormwater runoff from impervious surfaces. The continued growth in population density makes the looming urban water crisis a threat to economic prosperity. Fortunately, cost-effective solutions have been demonstrated for restoring healthy streams and neighborhoods. Originally developed for ecosystem management, GIS technology is proving to be a valuable tool for applying economic management at an ecosystem scale.

Bellevue, Washington, is presented as a case study that combines many elements of free market ecosystem management. Bellevue's boundaries closely coincide with those of the natural watershed. Unfortunately, the great majority of communities do not coincide with watershed boundaries, greatly complicating resource management. This is exemplified by a case study in Atlanta, Georgia, where GIS technology is applied at the community level to create watershed management units through better-defined property rights. The aim is to assist communities in a region with severe water problems to develop the mapping, data collection, and educational tools necessary for coordinating ecosystem management among multiple political jurisdictions within a demonstration watershed area.

Two approaches for improving urban water quality are presented here: community-based watershed protection and a new institutional arrangement known as a user-fee-based stormwater utility. Although the two approaches may at first seem unrelated, they, in fact, follow a natural progression from grassroots activism and public education to implementation through long-term dedicated funding.

FOUR PRINCIPLES FOR A FREE MARKET ENVIRONMENTAL AGENDA

As an alternative to the command-and-control approach, Anderson (1995) proposed three general principles for a new free market environmental agenda, along with a fourth principle that applies to water resources. These are as follows:

1. Secure and tradable property rights encourage cooperation and resource stewardship. Many water resource problems occur because natural watersheds are nearly always divided between multiple political jurisdictions with conflicting property rules and management goals. Competition for common-pool water resources places them up for grabs, where access is determined by political clout rather than market conditions.
2. Polluters should be liable for harms they cause others. Urban and suburban development is associated with an increase in impervious surface area from roads, rooftops, and parking areas. Scientific studies provide overwhelming evidence that relatively small increases in impervious surface area within a watershed result in dramatic declines in water quality from altered hydrology and polluted stormwater runoff. Property owners should pay a fee proportional to their impact, to be used in controlling runoff and meeting agreed-upon standards within a watershed.
3. Incentives matter. PERC promotes charging market rates in the arid West, for federally provided water in order to encourage conservation and increase instream flows. In the humid East and Pacific Northwest, where urban stormwater causes floods and damages aquatic health, this chapter argues for charging impact fees on impervious surfaces. Owners of wetlands and vegetated stream buffers, which create a net benefit to ecosystem health, should be rewarded by being exempt from impact fees. Where the ecosystem value of land exceeds its developed value or where the ecosystem is flood prone, impact fees should be used to purchase the land from willing sellers and restore it to its natural condition. This market-based approach contrasts sharply with the current policy of subsidizing uneconomic development through federal flood insurance and disaster payments.
4. Expand and improve the national system of water quality monitoring. Economists and ecologists agree that accurate information is essential for informed decision making. Current state surveys are incomplete and use varying methods based primarily on chemical grab samples that can miss episodic pollution. Annual data collected using standardized methods, including surveys of biological integrity, are needed on each U.S. river basin. Biological surveys provide a more effective and affordable

way to monitor the overall health of a watershed than expensive, continuous chemical monitoring. A water quality database meeting these requirements would allow government and citizens to determine whether local watershed standards are being met.

IMPACT OF URBANIZATION ON WATER QUALITY

Urbanization is having a damaging impact on water quality out of proportion to its actual rate of growth because the impervious surface area is reaching a biological threshold of 10 percent. A growing body of scientific evidence indicates that there is a direct link between impervious cover and stream health (Olympia Public Works Department 1995). Impervious cover in a watershed is a good indicator of the overall health of streams that feed rivers and lakes. Where impervious cover exceeds 5 percent, stream health begins to decline in some regions (May et al. 1997). With more than 10 percent impervious cover, stream health is biologically impaired. At 25 percent impervious cover, streams are nonsupporting.[1] Nutrient loading[2] and other types of pollution also increase in proportion to the impervious surface area in a watershed (Center for Watershed Protection 1998).

Urban development, which is accompanied by marked increases in impervious surface, now covers 5 percent of the land area in the United States. East of the Mississippi, the median amount of development—the midpoint when the states are ranked by their degree of development—is almost 10 percent (fig. 4.1). At the low end of the scale, residential development on es-

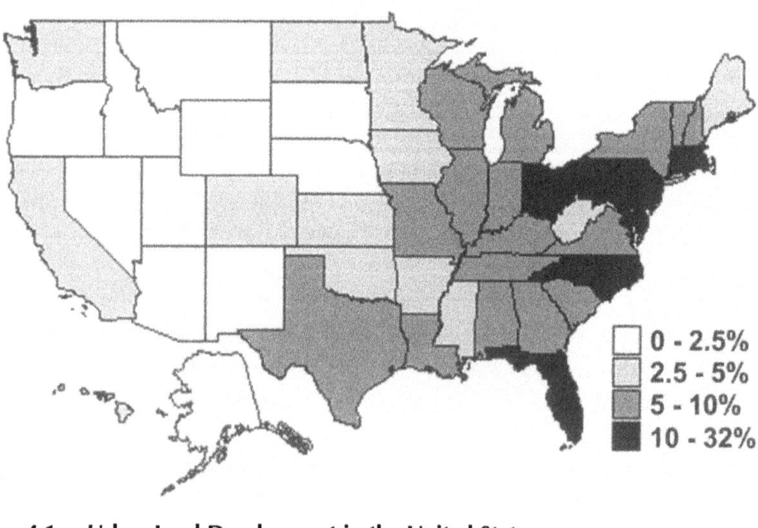

Figure 4.1. Urban Land Development in the United States
Source: Table 2 of Staley (1999, 11).

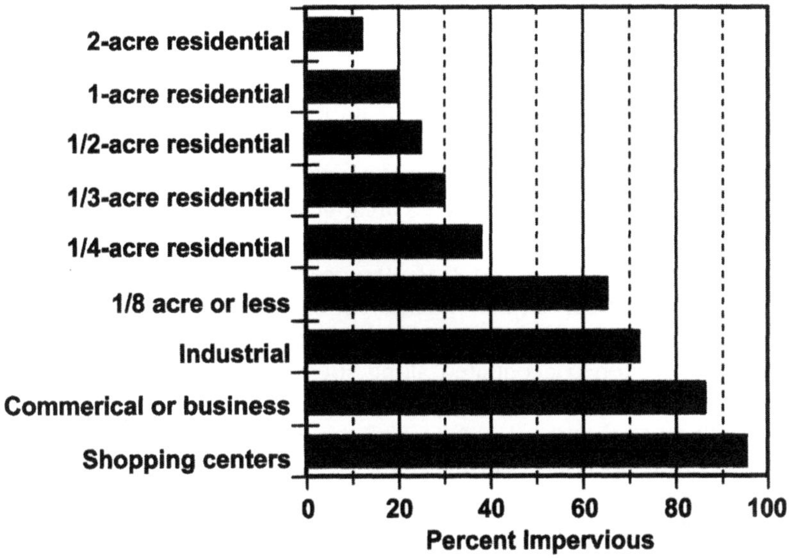

Figure 4.2. Impervious Cover as a Function of Land Use
Source: Arnold and Gibbons (1996).

tate lots covers 12 to 20 percent of the land with impervious surfaces. At the high end of the scale, shopping center development results in more than 90 percent imperviousness (Arnold and Gibbons 1996). See figure 4.2.

The rapid runoff of stormwater into urban streams erodes stream channels, causing them to deepen and widen, thus destroying aquatic habitat. Stream channel enlargement damages urban infrastructure by undermining bridges and sewer lines, collapsing structures, and eroding property. Increased runoff from impervious surfaces also results in an increase in flood damage. Sediment erosion during the construction phase of development and later from enlarging stream channels smothers aquatic life and accumulates downstream in water supply and flood control reservoirs. By increasing surface runoff and blocking the recharge of groundwater, impervious surfaces can cause small streams to dry up during droughts. The total cost of disruption to the hydrological cycle and resulting damage to urban infrastructure has not been calculated, but it is enormous.

An indication of how extensively development patterns are conflicting with nature can be gained from Mileti (1999). The study by a team of 132 academics and officials estimated damage from severe storms between 1975 and 1994 at $400 billion, not including indirect costs such as lost business or environmental harm. Seven of the ten most damaging storms, adjusted for inflation, occurred during the last five years of the twenty-year study period, indicating that damage is increasing. The report blames human development patterns for the increasing damage. One example was the devastating Midwest flood that destroyed 50,000 homes in 1993.

FRAGMENTED TRADITIONAL MANAGEMENT

Traditional stormwater management treats stormwater as a waste product of urban development.[3] The increase in impermeable area results in an increased volume of runoff that is rapidly conveyed by storm drains into streams. The development-induced surge of runoff from each storm then erodes stream channels draining the area, gradually enlarging them and depositing sediment downstream (Leopold 1994, 165). Over a period of twenty years, the effect is dramatic. At 25 to 30 percent impervious cover that is typical of many suburbs, stream channels experience two to five times stream channel enlargement (Center for Watershed Protection 1998). Because development is often concentrated along stream corridors, channel enlargement results in extensive property damage.

In addition to the harmful impact of impervious surfaces on hydrology, drainage networks collect pollution and discharge it into streams. Between rainfalls, pollutants settle and accumulate on impervious surfaces. When rain falls, pollutants—particulates from air pollution, motor oil, pesticides, fertilizers, heavy metals, pet feces, and others—that settle on road surfaces form a first flush of contamination that washes into streams and lakes. With scouring, erosion, and pollution following every rain, stream channels become biologically nonsupporting extensions of the storm sewer system and therefore require intensive treatment of water withdrawals for miles downstream.

Responsibility for stormwater management is generally dispersed among various government agencies and departments, creating conflict and regulatory overlap. At the federal level, water resources are addressed by a multitude of agencies including the U.S. Army Corps of Engineers, Federal Energy Regulatory Commission, Department of Transportation, Bureau of Reclamation, Natural Resource Conservation Service, Environmental Protection Agency (EPA), and Fish and Wildlife Service. At the municipal level, water management is generally divided between separate departments for drinking water, roads and drainage, sewers, and parks and recreation, giving stormwater management stepchild status to roads and development (Cyre 1999). Stormwater programs typically suffer from inadequate funding, forcing managers to react to problems such as flooding with short-term piecemeal solutions. Drainage systems are rarely built with the capacity to handle runoff from anticipated development. All too often flooding results when pavement is laid without providing for the control of additional stormwater. When local jurisdictions are forced to control stormwater that is polluting the supplies of other communities downstream, they are likely to respond as did the city of Atlanta, by building expensive end-of-the-pipe treatment plants.

Because political jurisdictions rarely coincide with watershed boundaries, water resource management is further fragmented between neighboring jurisdictions. When North America was settled, waterways served as the main avenues of commerce. In order to provide equal access and defensible boundaries, dividing lines for states, counties, and municipalities were drawn down the middle of rivers, streams, and lakes. Over time the practice of dividing surface waters between competing jurisdictions likely encouraged depletion, pollution, and waste. Efforts to resolve confrontations in court have led to protracted legal battles known regionally as water wars.

BELLEVUE PIONEERS FREE MARKET APPROACH TO ECOSYSTEM MANAGEMENT

An alternative, cooperative approach is watershed-based ecosystem management. This approach recognizes that a healthy human economy depends upon the services of nature for recycling air and water, stabilizing climate, and controlling pests, pollination, and other vital ecological processes. Each watershed is a unique ecosystem surrounded by a rim of high land and draining to a body of surface water. The interdependence of watersheds with headwater tributaries feeding creeks, rivers, lakes, and estuaries makes watersheds the appropriate geographical unit for the cooperative management of water resources. After all, political jurisdictions seldom coincide with watershed boundaries.

Bellevue, Washington, has protected its urban streams and environmental quality.[4] This progressive community is a suburb of Seattle, located between the shores of Lake Washington and Lake Sammamish.[5] The watersheds of the small headwater streams draining the city are almost wholly contained within the city's jurisdiction, greatly reducing the potential for management conflicts with neighboring jurisdictions. In addition to rediscovering the link between watershed protection and a healthy community, the city developed an incentive-based fee structure for maintaining the integrity of the system over time, making Bellevue a pioneer in free market, watershed-based ecosystem management.

Bellevue's program was the result of citizen concern over the impacts of rapid urbanization. The initial impetus for change came from residents who complained to the city council that the pace of development was destroying the very qualities that had attracted them to the area. The city council appointed several of the citizens to a committee to examine the problem. Initially this committee dealt with erosion and sedimentation resulting from development practices that were damaging stream channels used by spawning salmon. Recognizing that stormwater was causing the erosion, the committee recommended that the city establish an erosion control ordinance and a stormwater management program. At the same time, the state passed an amendment to existing legislation permitting stormwater management to be funded using a system of cost-based user fees.[6] Acting on the committee's recommendation, the Bellevue City Council established a stormwater utility in 1974 as a division of the Department of Public Works. The mission of the drainage utility was to "manage the storm and surface water system of Bellevue, to maintain a hydrologic balance, to prevent property damage, and to protect water quality for the safety and enjoyment of citizens and the preservation and enhancement of wildlife habitat" (Water Environment Federation 1994, 45).

GIS Database Helps Secure Property Rights

Soon after its formation, the Bellevue Utilities Department targeted land for purchase where its value for controlling floods, maintaining water quality, and providing wildlife habitat (ecosystem value) was high and the development value was relatively low. Over a twenty-year period, the stormwater utility has acquired property rights to more than 1,000 acres of wetlands, riparian corridors, and streams. Bellevue also has an extensive streamflow control system consisting of 10 main electronically controlled flow structures, 400 neighborhood runoff control facilities, and 950 privately owned flow-control systems (Bellevue Utilities Department 2000).

Although a considerable network of land throughout the city is managed for water quality protection, it also serves the citizens in other ways. Some sites flood only occasionally and are used for walking trails, bicycle paths, and recreation fields; others are visited by residents who enjoy watching wildlife. One frequently visited site is a wetland that is leased to a private operator as a blueberry farm. By minimizing pesticide applications, the wetland does double duty by growing berries and enhancing water quality. The site is particularly popular because residents can pay a small fee and pick their own blueberries or buy them by the quart.

Paper maps and records initially used for managing the utility soon proved inadequate for the complexity of the task. A key element of the stormwater utility concept is the development of a database that enables managers to overlay complex features of the built and natural environments. GIS stores layers of data on topography, vegetation, hydrology, water quality, geology, property parcels, roads, sewer lines, and drainage systems, and registers the data on a common cartographic coordinate system. Technology now makes it much easier to protect water quality by working with nature.

Pollution and excessive stormwater that flows from one property onto another can be considered a form of trespass. An extensive monitoring program operated by the utility makes it possible to regularly measure stream health and detect damaging flows and pollution. Because impervious surfaces have the greatest impact on stream health, most enforcement action is directed at prevention, thereby avoiding costly lawsuits. Building codes require onsite stormwater controls that reduce runoff and nonpoint pollution.

In addition to assigning responsibility for pollution, an important contributor to Bellevue's success in avoiding a tragedy of the commons for its streams and lakes is the sense of pride in ownership expressed by city residents. Several programs have reinforced community property rights.

1. Published watershed maps allow residents to identify the watershed they live in and to learn where their neighborhood streams flow. Just as road maps allow a person to navigate by car in an unfamiliar area, watershed maps can make the natural environment more familiar.
2. Visitors to Bellevue immediately notice that city streams are identified by signs just like city streets. Residents who get to know their watershed are more likely to develop an interest in protecting their streams.
3. A popular Adopt-a-Stream program, known locally as the Stream Team, takes children and adults on regular watershed walks. Once citizens get to know their streams and keep them healthy, the streams are viewed as a resource instead of a place to dispose of waste.

GIS is a tool that helps define property rights for complex natural systems. Unlike land lots with fixed boundaries, streams and rivers change over time due to variations in flow and sediment load. Stream ecosystems can be better defined using GIS technology because they are a blend of aquatic and terrestrial environments. During alternating periods of heavy rainfall and drought, a stream's wetted channel expands and contracts. A stream's vegetated border (or riparian zone) controls erosion, moderates temperature, provides habitat, and supplies organic debris that forms the base of the aquatic food chain. An intact riparian zone helps absorb pollutants before they can be carried downstream. The complex, dynamic character of the natural environment is more easily managed using GIS technology.

User Fees Apply the Polluter Pays Principle

Bellevue's program is innovative because its user-fee stormwater utility charges property owners in proportion to the impact of development on the health of the natural hydrologic cycle. Like other systems in nature, the hydrologic cycle is dynamic, adapting to change through feedback loops that work to restore equilibrium. The cost-based, user-fee rate structure establishes a feedback loop from impacts on the natural system, as measured through impervious surface area and uncontrolled runoff. Performance is measured by evaluating stormwater damage to property, the health of the aquatic system, and adjusting fees accordingly. Bellevue's desirability as a community is evidenced by its high property values.

For billing purposes, managers use the GIS system to calculate impervious surface area from aerial photographs overlaid with individual property parcels. The service charge is based on a graduated scale for the size of the property measured in square feet, and the amount of development determined by the percent of impervious surface. A heavily developed property is assessed about three times more than an undeveloped property of the same size; there is no charge for wetlands.[7] A typical residential homeowner pays about $100 per year.[8] As the city improves its GIS mapping and data retrieval system, it plans to refine the graduated fee structure to better reflect impacts. Unlike property taxes, which generally exempt government and institutional property, user fees apply to all property regardless of ownership.

Using the collected fees, the stormwater utility began an aggressive maintenance program to catch up with neglected repairs and acquired key properties deemed important for the protection of water quality and wildlife habitat. Public information programs featured previously unnoticed streams winding through the town. A demonstration project was begun featuring stream protection in cooperation with local citizens and state agencies. Over time hundreds of small-scale neighborhood facilities and several regional control facilities were constructed to moderate and cleanse stormwater runoff from impervious surfaces. These facilities function as ponds, wetlands, or groundwater recharge areas. User fees equitably assign the cost of stormwater services, where customers causing the greatest impact pay the highest fee. Developers can lower their fee by limiting impervious surface area or controlling runoff through more efficient designs.

The Stream Team provides technical assistance to property owners who face erosion or want to enhance their stream habitat. In many cities, property owners pour concrete or dump rocks on eroding streambanks, a practice that often slows erosion at the site but accelerates erosion for other property owners downstream. This is one reason Bellevue requires owners to purchase a permit before disturbing stream habitat. As an inducement to encourage streamside property owners to control erosion by healing streambanks, the owners can have permit fees waived when they work with the Stream Team.

Water Quality Monitoring and the Public Information Program

User fees support a regular program of stream monitoring using chemical and biological testing. Monitoring results obtained in the field are tied to specific map locations using the GIS system. A serious problem faced by citizens and government is gaining access to the volumes of data collected on water quality and other measures of environmental health. GIS systems have the power to store and display data in a user-friendly way by linking it to familiar locations on a map.

The interactive nature of the GIS database makes it a valuable tool for education. As the environmental orientation of Bellevue's program grew, schools began incorporating stream monitoring into their curriculum of study. Bellevue has a full-time Adopt-a-Stream program that shares data between the schools and the utility department. One of the innovative features of the program is the use of maps that are printed from the GIS system and used for "stream walks" by science classes. Stream walks are field trips that give students the opportunity to observe the health of their neighborhood streams up close under the guidance of teachers and involved parents. Appropriate data collected during the walks is then entered into the GIS database, increasing its value as a practical resource. Because the system contains current information that is constantly being updated, it has become the central data repository for government and citizens.

Integrated Utility Applies the Pollution Prevention Principle

Since its founding, the Bellevue program has evolved to implement a watershed approach combining controls on runoff from impervious surfaces with protection of the natural drainage system of streams and wetlands. Maintenance practices have been altered to emphasize prevention. Despite the urban character of the community, water quality remains high enough to support several species of salmon and trout.[9] Recent storms have proven the value of stormwater controls, stream protection, and wetlands in preventing property damage from flooding. The stormwater utility was eventually elevated to department status and is now cooperating with surrounding King County to apply a watershed approach to comprehensive water resource management including water supply, sanitary sewage, stormwater drainage, and wildlife protection. The user-fee-funded utility provides functional recognition of stormwater as a resource. Under the concept of an integrated utility, responsibility is focused in a single department. Increased dedicated funding tied to impacts provides for comprehensive preventive programs, making sustainable development economically viable (Cyre 1999).

Evaluation of Bellevue's Free Market Ecosystem Approach

Despite one of the highest service fees in the nation, Bellevue's user-fee-funded integrated utility enjoys a high level of community support. In 1998 Bellevue hired a private contractor to conduct a performance survey of its utility customers (GMA Research Corp 1998). The ratings were favorable. The highest overall complaint with utility service (including water, sewer, garbage/recycling, drainage, and wildlife) was 9 percent of respondents saying fees were too high. Interestingly, when respondents were asked to rate the protection of fish and other wildlife, 44 percent said the utility should do more even if that meant higher bills. The Bellevue Stormwater Utility was rated as good or excellent by 66 percent of respondents, neutral or don't know by 23 percent, and poor or very poor by 12 percent.

ECOSYSTEM MANAGEMENT IN ATLANTA'S POLITICALLY FRAGMENTED WATERSHEDS

Bellevue's location was ideal for the establishment of a free market approach to watershed protection because the city's jurisdiction closely coincides with the natural watersheds that drain it. Unfortunately, most communities are not in such a fortunate circumstance. They have to deal with multiple political jurisdictions at the municipal level in addition to those at the county, state, and federal levels. Because Bellevue's streams originate within the city, responsibility for monitoring and protecting is clearly the city's responsibility. In other regions of the nation with a more typical pattern of multiple local jurisdictions overlying watersheds, assigning responsibility for stormwater runoff and nonpoint pollution has proven to be devilishly difficult.

Atlanta's Emerging Water Crisis

A case study of Atlanta, Georgia, illustrates the severe water problems of one of the fastest-growing metropolitan areas in the nation.[10] Atlanta has no geographic barriers to expansion; however, a finite supply of clean water is looming as a barrier to growth. Like most U.S. metropolitan areas, political fragmentation of watersheds greatly complicates effective water management. Atlanta is at the center of a bitter war with neighboring states over water rights. The protracted legal battle known regionally as the "water wars" began over Atlanta's demand for additional water from rivers shared with Alabama and Florida.

The battle began in the drought years of the 1980s when Atlanta faced serious water shortages, leading to conflicts over the allocation of limited supplies from the Chattahoochee and other rivers. The conflicts came to a head in 1990 when the state of Alabama sued the U.S. Army Corps of Engineers in federal court to block the allocation of additional water to municipalities within the metro Atlanta region. When Florida joined the dispute, the press dubbed it the Tri-State Water Wars. The suit was prompted by a plan to double metro Atlanta's permitted withdrawals by 2010 (Montgomery 1991). The parties are continuing to negotiate in order to avoid having the case go to the Supreme Court. The water wars suit reveals the difficulties of resolving resource conflicts through adversarial means.

A series of legal battles have also been fought over deteriorating water quality. A case with great implications for the rest of the nation was decided in 1996. The court ruled in favor of the Georgia Center for Law in the Public Interest, finding that Georgia and the EPA had failed to enforce the total maximum daily load (TMDL) requirements of the Clean Water Act. A TMDL is the limit on the amount of pollution a water body can receive without violating clean water standards. Under oversight by the EPA, Georgia must develop

TMDLs for all impaired waters in the state by 2005. The first watershed to be addressed contains the Chattahoochee River, which supplies water to and receives waste from metro Atlanta.

Similar lawsuits have now been filed in at least thirty states, requiring the EPA to enforce TMDL provisions of the Clean Water Act requiring that unhealthy water bodies be identified and cleaned up. In sixteen states, the EPA is under court order to establish schedules for cleanup if the states do not do so themselves (EPA 2000). In August 1999, the EPA announced a proposed rule requiring for the first time that all states take a watershed approach to identify impaired waters and submit a schedule for cleaning them up (*Atlanta Journal-Constitution* 1999).

Facing increasing pressure to stop polluting its rivers and streams, Atlanta began the Metro Atlanta Urban Watersheds Initiative (MAUWI) in early 1996. The goals of the initiative were to (1) determine the current conditions of creeks and rivers; (2) identify the source and effects of pollutants on the health of waterways, and; (3) devise a plan for improving water quality. A community-based committee of watershed stakeholders was appointed to guide the process.

In late 1998, consultants performing the MAUWI studies issued their report containing the following conclusions: (1) almost all of Atlanta's streams were moderately to severely impaired; (2) the main cause of both biotic and nonbiotic stream impairment was altered hydrology resulting from high levels of impervious surface; and (3) stormwater contributed approximately three-quarters of the pollutant loading in Atlanta streams (CH2M Hill 1998). The initiative also reported estimates to the city of the cost to improve the health of Atlanta's streams and initial discussions on the concept of a stormwater utility. The citizen-based steering committee recommended the watershed program be continued, but funding was not renewed and the program was terminated.

Like most areas of the country, watersheds in metro Atlanta are divided among multiple local political jurisdictions. The Atlanta Regional Commission predicts that a water crisis will soon appear if local governments within metro Atlanta do not take steps to address water management on a regional basis (Brownlow 1999). Unless there is cooperation at a local level, the EPA will be required to impose sanctions forcing a cleanup.

Scientific Panel Recommends Watershed Demonstration Sites

Recognizing the need for selecting demonstration sites to showcase watershed protection, the ILSI Risk Science Institute, funded by the Atlanta Community Foundation, convened an independent science-based panel of nationally recognized experts. The panel held a three-day conference, made field trips, and recommended four sites for demonstrating a watershed approach to solving Atlanta's water problems. The ILSI panel hoped their report would be a starting point for the development of community-based watershed management plans (ILSI Risk Science Institute 1998).

A valuable tool used in the selection process was a GIS database developed for the MAUWI. During the ILSI conference, the expert panel used the GIS system to project maps of area watersheds. The maps were then overlaid with other pertinent information on the natural and built environment, enabling the participants to select candidate sites that were then visited during field trips. A major shortcoming of the system was that maps lacked sufficient detail at the local level. For example, determining property ownership patterns along streams was a time-consuming task because the records were not readily available. Another problem was that features of the natural environment such as small streams were either missing or inaccurate.

At the time, the author was a community representative serving as chairman of MAUWI. Before funding for MAUWI ran out, the steering committee voted to continue development of the GIS system as a priority item. One of the consulting firms working on MAUWI was W. L. Jorden and Associates. W. L. Jorden worked with the author on a pro bono basis to further develop the GIS system for one of the four watersheds identified by the ILSI panel as a potential demonstration site. Using Arc Info software produced by Environmental Systems Research Institute (ESRI), a highly detailed map was created using data layers for the built and natural environments. The layers included property parcels, sewer lines, hydrology, impervious surface area, and water quality data. Photographs were also entered into the GIS system using Arc View software, creating a compelling visual presentation. The GIS-based presentation was shown to local communities as a useful tool for watershed protection.

PROPERTY RIGHTS APPROACH TO COMMUNITY-BASED WATERSHED PROTECTION

Paideia School is a private, nonprofit school located in the Druid Hills section of Atlanta, Georgia, and provides a quality education for 800 students ages three to eighteen. The school received a donation of 4.2 acres of undeveloped land that is traversed by the Lullwater Fork of Peavine Creek.[11] The school is located four miles from the center of Atlanta within the drainage basin of the Appalachicola-Chattahoochee-Flint River basin, which spans three states. Lullwater Fork is a neighborhood stream draining a subwatershed of 1,000 acres.

Peavine Creek is a tributary of the Chattahoochee River that supplies drinking water to more than half of Georgia's population. Atlanta's rapid growth and its location near the headwaters of its principal supply river threaten its future economic health. A study by the MAUWI determined that almost all of the urban streams emptying into the Chattahoochee River are in a poor state of health. The main cause of impairment is stormwater runoff from impervious surfaces. The proposed solution was watershed protection beginning in the headwaters. A second cause of impairment is leaking sewer lines that will need to be identified and repaired. An independent panel of scientific/technical experts convened by the ILSI Risk Science Institute selected the Lullwater Fork as a proposed demonstration site.[12]

In studying how to use its land, Paideia School learned about the importance of watershed management and its effect on the school's ownership of the property. The banks of Lullwater Fork are eroding and the stream is badly polluted. Since the problems originate upstream, Paideia has worked to build partnerships with property owners in the headwaters. From this study and partnership building, a plan has developed to use the Lullwater Fork as a demonstration project for innovative community-based watershed planning. Paideia School wishes to use its property as an outdoor classroom. The demonstration project will provide educational opportunities for students at Paideia and other local schools. More important, the concepts developed by Paideia and its partners will be transferable to other property owners and communities as they address their own watershed management issues. The GIS system will be used to enter information from stream monitoring and as a tool for developing watershed plans. The project consists of the following components:

- A computerized GIS with map layers including hydrology, water quality data, sewer system, and property ownership.
- Monitoring of water quality through streamflow gauges and professional biological surveys. An effort will be made to improve the accuracy of Adopt-a-Stream data through training and comparison with professional samples.
- Education on watershed protection following a property rights approach by using Paideia School's property as an outdoor classroom; siting three outdoor exhibits in public places near the creeks; and holding meetings to involve residents, inform public officials, and notify media.
- A watershed master plan developed in conjunction with the Center for Watershed Protection. A key goal will be rehabilitation of Lullwater Fork as measured by an index of biological integrity.
- Raising additional funds to rehabilitate Lullwater Fork beginning in the headwaters.

Watershed Planning Approach

Until now, government has addressed these problems in isolation, one at a time, through large public works projects. This project takes a watershed approach that combines ecology and economics to address multiple environmental problems, primarily by restoring the natural ability of trees and streams to filter the air and water. The plan relies on pollution prevention at the source, as opposed to the current approach, which emphasizes the construction of treatment plants to clean up water pollution after it occurs. Because natural systems tend to be self-sustaining, long-term costs are projected to be much lower than maintenance-intensive treatment plants.

Treatment plants make bad neighbors and depress property values. They are perceived as being ugly, noisy, smelly, and dangerous. A history of effluent discharges and the increased use of chlorine, which is toxic to fish and other aquatic life, have confirmed neighbors' fears. The grant partners share the view that preventing pollution and restoring the natural assimilative capacity of streams[13] will result in ecologically healthy neighborhoods that are safer, more desirable places to live. These same qualities will support higher property values.

The proposed watershed approach is community based because it is the result of grassroots efforts to improve neighborhood quality. The grant recipients are using the funds to form partnerships with the different levels of government and upstream neighborhoods. These partnerships will be essential for sharing data and implementing proposed management changes in the watershed.

An essential component of the grant is funding for monitoring through the Adopt-a-Stream program. Many urban children are growing up with absolutely no connection to their environment. They expect government to provide them with clean water, but they have no concept that clean water can come from any source other than a pipe. Adopt-a-Stream teaches children how to monitor the health of streams that drain their neighborhoods. Field trips allow children to compare their neighborhood stream with a healthy stream. In the process, many begin to understand that clean water is their responsibility.

Peavine Creek Selected for Two Demonstration Sites

Members of the ILSI panel identified two portions of the Peavine Creek watershed as excellent sites for demonstrating sub-basin watershed management. The first site is Fernbank Fork. There, the Lake Claire Neighbors Association is working with The Nature Conservancy to create and implement a watershed management plan funded by the Walker foundation. Fernbank Fork was chosen because the subwatershed contains a forest preserve, Fernbank Forest, that supports a small spring-fed tributary that is still in excellent health. This is a rare, scientifically valuable reference stream needed as an example for restoring and restocking other impaired streams; however, the small stream is threatened. The existing grant supports a community effort to improve the health of the surrounding watershed and reduce habitat fragmentation.

The second site is Lullwater Fork. It was selected by the ILSI panel as a demonstration site because it is a highly degraded headwater stream that can be rehabilitated to a healthier condition for the least cost. The two streams are part of the same ecosystem and flow together to form Peavine Creek. The headwaters of the Lullwater Fork are located to the south and west of the Paideia land, and significant portions are in either Freedom Park or other Atlanta city parks. The availability of public land in the headwaters of Lullwater Fork greatly lowers the cost of rehabilitation compared with other potential remediation sites. Downstream from the Paideia property, Lullwater Fork flows through a vegetated riparian corridor, a legacy of the design by the famous landscape architect, Frederick Law Olmsted Sr., for the Druid Hills neighborhood.

A portion of the vegetated stream corridor is privately owned and maintained by the Lullwater Garden Club as a nature park. The park, called the Lullwater Conservation Garden, is open to the public. This beautiful section of Atlanta is listed on the National Register of Historic Places (across the street from the conservation garden is the house used as the set for filming the Hollywood movie *Driving Miss Daisy*). Streambank erosion caused by stormwater runoff is threatening many of the beautiful, mature hardwood trees lining this portion of the Lullwater Fork. An immediate concern of the club that owns the park is the dramatic increase in maintenance costs due to stream erosion.

Funding Sources Available for Remediation

Implementation of the watershed plan developed through this grant will require larger-scale funding. There is reason to be encouraged that funding will be available to implement a watershed demonstration project. Watershed protection has been successful in small-scale projects in other parts of the United States. The most successful example of urban watershed protection is in Bellevue, Washington, a community that turned a community-based effort into a city-wide approach by funding watershed protection through a user-fee-based stormwater utility.

A user-fee-funded stormwater utility bases its charges on the amount of impervious surface area. Stormwater runoff from impervious (primarily paved) surfaces has been shown to be responsible for 75 percent of the pollutant loading in Atlanta streams draining into the Chattahoochee River. It is also responsible for rapid runoff that causes increased flooding and damage to infrastructure such as sewer lines, bridges, and storm drains, as well as to private property through increased streambank erosion. By avoiding such infrastructure damage, the burden on taxpayers is eventually lessened. Utility funds could be used for watershed protection efforts such as restored wetlands that reduce flooding and provide aquatic habitat, instead of hard-engineered projects such as Atlanta's recently constructed combined sewer overflow/stormwater treatment plant in Piedmont Park, which discharges stormwater into Clear Creek after disinfecting it with chlorine that is toxic to fish and other aquatic life.

In January 1999, Atlanta approved a stormwater utility and began collecting fees. Unlike Bellevue, which charges a user fee based on impervious surface area, Atlanta's stormwater utility calculated its charge based on the square footage and use of each property. In September, Atlanta's stormwater utility was overturned in court because it is similar to a tax used to raise money for general purposes. The city did not satisfy the court that the funds were dedicated to stormwater and water quality improvements (Campos and Hairston 1999). A successful user-fee-based stormwater utility in the city of Griffin, just south of Atlanta, has established a legal precedent and has demonstrated that the concept will work in Georgia if the user fee is applied to dedicated funding of stormwater improvements.

Bellevue's successful user-fee-based stormwater utility is popular with rate payers because it evolved out of a community-based effort that demonstrated successful results as it developed. It also uses GIS mapping for management and determining fees. This grant could help initiate a similar success story in Atlanta and result in a long-term source of funding that is proportional to the impact causing water quality impairment.

The small headwater tributaries that feed Lullwater Fork are buried in culverts or confined in concrete ditches. These structures are collapsing. A project by DeKalb County in nearby Longdale Park showed that when costs for labor and materials are taken into account, "daylighting a stream," or restoring a naturally flowing stream to the surface, can be cheaper than burying it in a new culvert. Following a 1994 public bond referendum in which Atlanta raised funds for urgent drainage repairs, $250,000 was dedicated for drainage improvements in Candler Park, site of collapsing stream culverts. The Candler Park Community Association hopes to raise additional funds and use the watershed plan developed through this project for controlling stormwater runoff and restoring the stream.

IMPLICATIONS FOR THE FUTURE

Until recently, detailed information relating individual property parcels to the features of urban watersheds was practically nonexistent. The creation of property rights for the environment is an essential step in forestalling a tragedy of the commons for our water resources. Three revolutions in technology now make a property rights approach to environmental protection affordable for communities throughout the United States.

The first revolution is in GIS technology that integrates ecosystem and economic data based on spatial location. GIS enables air, water, and biological habitats, entities that cannot be fenced like real property, to be better defined. Originally developed for ecosystem management at the scale of river basins, GIS technology is being successfully applied by stormwater utilities to local watershed management.

The second revolution is in remote imaging and computer processing that is bringing down the cost of obtaining data. Ecosystem management requires information on a wide range of scales from multistate river basins to neighborhood streams. Despite the early promise of remote sensing with the launch of the first Landsat satellite in 1972, its use was limited by the coarse resolution of the satellite image.[14] Although new satellites with improved resolution were launched during the 1970s and 1980s, they still could not provide usable data for most applications due to the sophisticated software and computer requirements unavailable to most users. The recent development of powerful GIS software and affordable personal computers is ushering in a new era in remote sensing analysis. The licensing of companies to launch and operate high-resolution imaging satellites and sell those images will make precise mapping of environmental impacts commonplace. Three companies have launched, or will soon launch, one-meter-resolution imaging satellites capable of worldwide coverage (Mercado 1999).

The third revolution in electronic communication will make detailed information on environmental health and costs available to the concerned citizen. As consumers of manufactured goods, we take for granted the availability of detailed product information on price and utility, yet this information is completely lacking for ecosystem services such as the natural recycling of clean air and water. Environmental defense is one of the first organizations to seize upon the power of information technology to spur environmental improvement. Their popular Website (www.scorecard.org) lists previously unpublished data on local area airborne contaminants. Watershed groups are beginning to benefit from the distribution of information and GIS maps over the web. A new open records law in Georgia addresses the refusal by some local governments to allow public access to maps in electronic form, by requiring them to provide the information at reasonable cost. Access to detailed information required for secure and tradable property rights in the environment is rapidly becoming available.

Hardin observed that as population becomes denser, the natural biological recycling processes become overloaded, calling for the creation or redefinition of property rights. Scientific studies linking increases in impervious surfaces, mainly roads and parking lots, to a rapid decline in biological health and water quality show that correcting the problem will require a fundamental shift in management. Community-based watershed protection offers an appealing strategy aimed at protecting water at its source. This strategy is often preferable to removing pollution once it enters a water supply. In practice this requires local, decentralized decision making. The EPA (1998, 7) has estimated that more than 3.5 million miles of streams and rivers exist in the United States. Given the enormity and complexity of the task, keeping our waters healthy will require a local effort. The recently created institution of the cost-based, user-fee-funded stormwater utility is suggested as an effective way of applying a property rights approach to watershed management at the local level.

NOTES

1. Nonsupporting streams are essentially conduits for stormwater that no longer support a healthy community of aquatic organisms. The channels are typically eroded and enlarged, and often are prone to greatly increased flooding.

2. Nutrient loading occurs when nutrients naturally found in healthy streams are increased due to contaminated runoff and other causes. Nutrient loading stimulates the growth of algae and other aquatic vegetation leading to eutrophication or accelerated aging of receiving bodies of water.

3. Less well recognized is the role stormwater drainage plays in water pollution from agricultural fields. Drained wetlands, floodplain fields, and irrigated fields are often underlain by a system of drainage pipes or drained by a network of ditches. These drainage systems discharge runoff that is loaded with fertilizers and pesticides directly into surface waters.

4. Site visit, Utilities Department, Bellevue, Washington, August 4–6, 1997.

5. Bellevue has a population of 110,000. The utility department has determined that most of the city exceeds 30 percent imperviousness.

6. The state of Washington provides for the formation of stormwater utilities funded through user fees rather than property taxes in the Municipal Utilities Act, chapters 35.67 and 36.94 RCW.

7. City of Bellevue Ordinance No. 5115.

8. Rick Watson, operations manager, Utilities Department, Environment Division, Bellevue, Washington, telephone interview, August 3, 1999. This is the portion of the total utilities bill attributable to stormwater.

9. Bellevue Utilities Department lists the following salmonid fish in city streams: salmon (chinook, coho, sockeye) and trout (cutthroat, rainbow, steelhead).

10. Metro Atlanta has a population of 3 million and receives approximately fifty inches of rainfall per year.

11. Local nomenclature regarding the stream is confusing. Some residents think of the stream as being separate and call it Lullwater Creek; others know that it is a tributary of Peavine Creek and call it the Lullwater Fork of Peavine Creek. To avoid confusion, and for consistency, throughout this chapter it will be referred to as Lullwater Fork.

12. The Alex C. Walker Educational and Charitable Foundation addresses issues related to the causes of economic imbalances from a free market perspective. Two organizations located within adjacent subwatersheds applied for and received $55,000 in funding from the foundation. Matching grants approximately doubled the

total amount applied to the projects. Both of the watersheds had been previously selected as demonstration sites by the ILSI Risk Science Institute panel. Because each watershed is unique, the funding requests addressed particular local needs. But both grants take a cooperative property rights approach to watershed protection. The grant to Paideia School is of particular interest from the perspective of property rights and education. The following plan for initiating a property rights approach to watershed protection is summarized from Paideia School's grant request to the Walker foundation.

13. The natural assimilative capacity of streams is the ability to absorb limited amounts of waste.

14. Multispectral images from space can be analyzed to reveal surface qualities including imperviousness, vegetation type, elevation, and so on.

REFERENCES

Anderson, Terry L., ed. 1995. Reinventing Environmentalism in the New Era. *PERC Policy Series*, PS-2. Bozeman, Mont.: Political Economy Research Center, February.

Arnold, Chester L., and James Gibbons. 1996. Impervious Surface Coverage: The Emergence of a Key Environmental Indicator. *Journal of the American Planning Association* 62(2): 247–58.

Atlanta Journal-Constitution. 1999. EPA Rule Puts More Burden on States to Clean Waterways. August 15.

Bellevue Utilities Department. 2000. *Bellevue's Efforts to Save Salmon.* Available: www.ci.bellevue.wa.us/utilities/surface/salmonefforts.htm. Cited May 10, 2000.

Brownlow, Rick. 1999. Intergovernmental Cooperation. Case study presented at sustainable development conference sponsored by the city of Atlanta, August 20.

Campos, Carlos, and Julie B. Hairston. 1999. Atlanta's Storm Fee Voided. *Atlanta Journal-Constitution*, September 29, 1.

Center for Watershed Protection. 1998. Watershed Protection Techniques. In *Rapid Watershed Planning Handbook: A Comprehensive Guide for Managing Urbanized Watersheds.* Ellicott City, Md., 1.18–1.22.

CH2M Hill. 1998. Impacts Assessment (East & West Watersheds): Metro Atlanta Urban Watersheds Initiative. Study prepared for Atlanta Department of Public Works, Division of Wastewater Services. September.

Cyre, Hector J. 1999. Stormwater Utilities. Presentation at Sustainable Development Conference sponsored by the city of Atlanta, August 20.

EPA, Office of Water. 1998. *The Quality of Our Nation's Water.* EPA 841-S-97-001. Washington, D.C., April.

———. 2000. Total Maximum Daily Load (TMDL) Program. January 31, 2000. Available: www.epa.gov/OWOW/tmdl/index.html. Cited October 23, 2000.

GMA Research Corp. 1998. Bellevue Utilities Performance Measurement Survey Summary Report. Prepared for Bellevue (WA) Department of Planning and Community Development. July.

Hardin, Garrett. 1968. The Tragedy of the Commons. *Science* 162: 1243–48.

ILSI Risk Science Institute. 1998. Mitigation of Urban Runoff Impacts on Atlanta Streams. Report prepared for the Atlanta Community Foundation, April 29.

Leopold, Luna B. 1994. *A View of the River.* Cambridge, Mass.: Harvard University Press.

May, Christopher W., Richard R. Horner, Jamers R. Karr, Brian W. Mar, and Eugene B. Welch. 1997. Effects of Urbanization on Small Streams in the Puget Sound Lowland Ecoregion. *Watershed Protection Techniques* 2(4): 483–94.

Mercado, Raul. 1999. Are You Ready for the Second Satellite Imagery Revolution? *Stormwater Fact Sheet.* Tallahassee: Florida Association of Stormwater. January.

Mileti, Dennis S. 1999. *Disasters by Design: A Reassessment of Natural Hazards in the United States.* Washington, D.C.: Joseph Henry Press.

Montgomery, Robert. 1991. *Living Waters: A Special Report on an American River.* Montgomery, Ala.: B.A.S.S., Inc.

Olympia Public Works Department. 1995. *Impervious Surface Reduction Study.* Olympia, Wash. May.

Staley, Samuel. 1999. *The Sprawling of America: In Defense of the Dynamic City.* Policy Study 251. Los Angeles: Reason Public Policy Institute.

Water Environment Federation. 1994. *User-Fee-Funded Stormwater Utilities.* Special publication. Alexandria, Va.

5

Technology and Property Rights in Fisheries Management

Daniel Huppert and Gunnar Knapp

The world's ocean and inland fisheries are of vast economic importance and potential, but under open-access regimes they operate inefficiently and often deplete the fish stocks. Fishermen can gain from rational conservation of fish stocks, but without individual property rights to the fish or collective fishing rules, each has little or no individual economic incentive to sacrifice current catch for long-term future harvests. Since H. Scott Gordon's (1954) initial explanation of the economics of common property fisheries, economists have considered the open-access fisheries a classic case of market failure. In a nutshell, the fish stocks are depleted, excessive capital and labor are devoted to fishing, the potential net social benefits (the resource rent) from fishing are dissipated, and the hurried and crowded fishery diminishes safety at sea and quality of fish delivered to markets. When central authorities or local communities exercise effective control over the aggregate level and patterns of harvesting to ensure better fish stock conservation, the resulting regulated open-access fishery may continue to squander resources in excessive costs and poorly timed fish marketing (Homans and Wilen 1997). This dismal economic performance can be blamed on the failure of the property rights/market mechanism or, more fundamentally, on the failure of social institutions and private contracting to create a better system.

The magnitude of these losses makes the resolution of the economic problems in fisheries a worthwhile pursuit. The Food and Agricultural Organization of the United Nations (FAO 1998) reports that overall world fish production increased from 20 million metric tons (mmt) in 1950 to 122 mmt in 1997. About 22 percent of this total comes from aquaculture, which operates with relatively well-defined individual property rights for rearing captive fish and shellfish. The aquaculture sector, which roughly doubled since 1990, produces large quantities of oysters, trout, carp, shrimp, and salmon. This is largely an Asian phenomenon, with China accounting for more than half of the world's total aquaculture production. The remaining 78 percent of world fish production are harvested from wild fish populations (table 5.1). According to the FAO (1998), exports of fish products in 1996 reached a value of US$52.5 billion, representing roughly 11 percent of the value of agricultural exports and about 1 percent of total merchandise trade.

Although some fish stocks are managed for sustained yield, many are heavily exploited, and others seriously depleted. Among the major ocean fish stocks monitored by the FAO (1998), 44 percent are reckoned to be fully exploited, 16 percent to be overfished, and 6 percent to be seriously depleted. The remaining one-third of the world's fish stocks, mostly low-value species in less accessible areas, is capable of sustaining some increase in harvest. In the developed world there have been some spectacular conservation failures, including the collapse of North Atlantic cod off Newfoundland, Canada, and several Pacific salmon populations listed as threatened with extinction in California, Oregon, and Washington. The depleted state of many fish stocks is not due entirely to fishing. Other contributing factors include the contamination of rivers and estuaries, and alteration of freshwater and nearshore marine fish habitats.

Economic losses associated with open-access overfishing are compounded by government-encouraged overinvestment in fishing fleets. Comparing gross market value of worldwide fish production to estimated costs, Christy (1997) finds a global fish deficit of $41 billion, fully one-third of the gross value. This is not predicted by even the most dire tragedy of the commons theory because fishing firms will not consistently spend more on operating and capital costs than they receive in sales revenue. According to Christy, the explanation for this fisheries deficit is that many nations commonly address economic distress in the fishing sector by providing subsidies of all sorts: direct payments, low-cost loans, technical assistance, gear loans, free harbor facilities, price supports, and marketing programs. These subsidies allow fisheries to continue operating even when overall costs exceed sales.

One practical response to the joint problems of overfishing and rent dissipation is the creation of more exclusive fishing rights, which focus the economic incentives of rights holders on conservation and encourage economically efficient harvesting and marketing. Rapid technological changes have

Table 5.1. World Fisheries Production, 1997 (million metric tons)

	Inland	*Marine*
Harvest from wild stocks	7.70	86.03
Aquaculture	17.13	11.14
Total	24.83	97.17

Source: FAO (1998).

facilitated the definition, exercise, and enforcement of these rights. This chapter examines the relationship between technological change and the evolution of rights-based systems of fisheries management.

CHALLENGES TO ESTABLISHMENT OF RIGHTS-BASED SYSTEMS OF FISHERIES MANAGEMENT

Fisheries management encompasses any of a broad set of customs, informal agreements, and formal rules for harvesting fish to conserve fish stocks and improve the benefits to people over time. The past century of fisheries management in northern Europe and North America has focused on the invention and application of rules to prevent overfishing in the face of increasing demand for fish and increasingly sophisticated technologies for locating and harvesting fish.

Until recently the trend in fisheries management was toward regulated open-access systems (Homans and Wilen 1997). Examples of fishing regulations include closures of fishing seasons and geographic areas, limitations on types and amounts of gear used by individuals or firms, and rules regarding the species and sizes of fish that may be caught or kept. Regulated open-access systems leave individual fishing firms to compete with each other to harvest the allowable catch, thus encouraging overcapitalization of fishing fleets and excessive fishing costs as diagnosed by a generation of natural resource economists (Crutchfield and Pontecorvo 1969; Gordon 1954; Scott 1955).

It was not until late in the twentieth century that property rights systems evolved in fisheries management. To appreciate why this is so—and why it is still difficult to establish exclusive harvest rights in the majority of fisheries—one must understand the unique technical, economic, and social milieu of ocean fisheries. These conditions limit the ability of fisheries management systems to develop either market-based exclusive rights systems or cooperative systems of collective rules for resource use.

Complexity and Inadequate Ecological Knowledge

A huge number and variety of fish and shellfish are harvested in the United States alone—more than 105 species had an ex-vessel value of greater than $1 million in 1998 (National Marine Fisheries Service [NMFS] 1999). The life cycles of these fish are vastly varied, as are the places in which they are harvested, the people who harvest them, and the methods used to harvest them.

Fish are mobile. Some migrate hundreds and thousands of miles across local and international boundary lines. Some live in different places at different stages of their life cycles. Fish of different ages and species swim together, making it difficult or impossible to harvest only those of a particular species or particular age. As a consequence, outside of aquaculture facilities, there is no analog to the rancher's barbed wire to assist in identifying who owns what fish or in controlling the movements of the fish.

We have only a rudimentary understanding of how marine ecosystems link commercially valuable fish to primary production in the ocean and to populations of predators and prey species. We have limited understanding of how harvests affect fish stocks and potential future harvests, not only of the fish we catch but also of other fish and nonfish species that are indirectly affected by harvests. We also have limited understanding of how fish are affected by other human interactions with the ecosystem or by factors beyond human control. As a result, even if we could accurately identify and confine specific, valuable fish, we would have difficulty controlling the indirect effects of harvests on the suite of interacting species.

All these factors make the optimal use of fisheries exceedingly complex and difficult in theory, and a daunting problem in practice regardless of the goals over which we attempt to optimize the system. Over time and with growing understanding of complex marine ecosystems, fishery managers' appreciation of these technical challenges has improved.

Social and Political Conditions

The social and political institutions controlling fisheries complicate the exercise of authority by fishery management authorities. Fishery management institutions should reflect the technical and spatial complexity of fish populations, but they often conform to political jurisdictions that poorly reflect ecological domains. We have evolved a complex, hierarchical structure of centralized and local authorities, and private and public decisions. It includes the United Nations Convention on the Law of the Sea and federal authority in the Exclusive Economic Zone (EEZ); the regional management bodies such as the federally sponsored Regional Fishery Management Councils in the United States; and more local public decision processes, such as state fish and game commissions. Private control is exercised over the actual commercial fishing, processing, and marketing processes. In addition, we have multinational treaties for managing Pacific halibut, Pacific salmon, tunas, and other species.

The fishing industry is a rich and heterogeneous organization. In the Bering Sea, small-scale independent firms sometimes fish alongside large ships owned by multinational fishing corporations, and onshore processing companies compete with offshore processors (factory trawlers and motherships) for shares of the catch.

The tradition of open-access fishing is solidly entrenched in culture and law. It is a common law doctrine that "the resources of the seas within U.S. jurisdiction belong to the public and that the government holds them in trust for the public." This "confers on government a continuing duty of supervision" and "reinforces concerns about the 'giveaway' of public resources to private interests" (National Research Council 1999, 39). Whereas free access to ocean fisheries was established when states had little ability or reason to control access, this legal status has become a social tradition. The notion that everyone has a right to go fishing was widely respected in law and practice until recently. For example, the Alaska State Constitution holds: "Wherever occurring in the natural state, fish, wildlife and waters are reserved to the people for common use" (Alaska constitution, art. VIII, sec. 3).

Given the complexity of the ecological-economic system, and the social, cultural, and legal obstacles to emergence of property rights in fisheries, it is not surprising that fisheries are still largely open-access resources regulated by government agencies. If, as suggested by Libecap (1989, 92), distributional conflicts hinder the adoption of more complete property rights, the complex economic and other interests in fisheries present a serious obstacle to exclusive harvest rights procedures.

EVOLUTION OF NATIONAL RIGHTS TO OCEAN TERRITORY AND FISH

Property rights in fisheries rest on a foundation of national and international jurisprudence. Without recognized national rights of jurisdiction over the coastal oceans and fish, nations cannot effectively recognize or enforce individual or collective rights to harvest, manage, or conserve fish. Thus the first step in establishing more exclusive rights to harvest fish is to formulate an international standard for dividing up rights to the ocean among interested nations.

The Three-Mile Limit for Territory and Fish

From roughly 1600 to 1782, a variety of standards for national sovereignty over coastal seas were proposed and claimed. Most nations claimed rights over areas within which coastal states could control ships via coastal batteries. Apparently the "cannon-shot rule" was translated into the three-mile rule by the U.S. Congress in the Neutrality Act of June 5, 1794, at the outbreak of war between Great Britain and France. The three-mile rule came to reflect the outer limits of territory for purposes of customs patrol, ship traffic, neutrality in time of war, and licensed fishery (O'Connell 1982, 125–31). The three-mile territorial sea reigned as the international standard for fisheries and other purposes until after World War II, when it was supplanted by the Truman Doctrine, which claimed U.S. sovereignty over fisheries and other resources out to twelve miles, while retaining a territorial sea of three miles.

Law of the Sea Treaty: Extending Fisheries Authority from 12 to 200 Miles

Property rights in fisheries and other ocean resources have evolved steadily since World War II. Burke (1994) describes a series of international conventions, treaties, and bilateral agreements that modify and clarify rights to marine resources, culminating in the 1982 United Nations Convention on the Law of the Sea (LOS). This international treaty assigns primary interest in resources in the 200-mile EEZs to the coastal nations. On the high seas outside the EEZs there remains a presumption of freedom of fishing. But this freedom is significantly truncated in specific instances by international treaties and United Nations resolutions. Examples include the U.S.–Canada Pacific Salmon Treaty, which allocates salmon catch between nations, and the North Pacific Anadromous Fish Commission, which prohibits Pacific salmon fishing on the high seas (Burke 1994, 177).

These changes in international and national law clarify the ownership of fish stocks, facilitate the implementation of fishing regulations, and create the opportunity for coastal states to define more exclusive harvest rights within their jurisdictions. Each nation has significant latitude to delegate more or less authority within its EEZ to subnational units.

EVOLUTION OF RIGHTS-BASED FISHERY MANAGEMENT SYSTEMS

The first modern wave of change to the prevailing open-access system involved closing access to the fishery via exclusive access rights, most commonly in the form of license limitation or limited entry.[1] As indicated in table 5.2, license limitation systems provide a harvest right in common with other license holders, but they give each licensee no specific right to a quantity of fish. Sometimes the right may be assigned to an identifiable group, for example, the assignment of rights to 50 percent of salmon harvests in usual and accustomed fishing grounds to treaty Indian tribes in Washington State.

In principle, the designation of license holders eliminates the open-access problem by formally creating a commons with named co-owners. This is a necessary but rarely sufficient condition to encourage users to organize an efficient harvesting system. So long as there are large numbers of licenses, and so long as each has an economic incentive to catch the fish before the others, harvesters will continue competing for shares of harvest through strenuous fishing that requires excessive capital and labor over short periods. But long-term high market prices for Alaska's salmon permits show that license limitation has some economizing effect, even in fisheries with hundreds of small-scale fishing firms (Huppert, Ellis, and Noble 1996). Moreover, when the number of license holders is small and the legal institutions support collective resource conservation or enhancement, the license holders may cooperate to improve economic payoffs (Higgins 1999).

The next move toward rights-based fishery management came in the form of harvest share rights, under which individuals or groups hold rights to harvest specific shares of the annual total quota. In theory, harvest share rights encourage more efficient harvesting and marketing in commercial fisheries (Moloney and Pearse 1979). Variously termed individual quotas (IQs), individual fishing quotas (IFQs), individual transferable quotas (ITQs), enterprise quotas, and quota shares, these quantitative harvest rights were used first in enclosed lakes and then in the oceans. Exclusive harvest rights systems were pioneered in Canada (Casey et al. 1995; Gardner 1989), New Zealand (Annala 1996; Boyd and Dewees 1992), Iceland (Arnason 1997), and Australia (Geen and Nayar 1989). Individual quota systems came to the United States in the early 1990s (National Research Council 1999).

Despite the fact that well-defined harvest rights promise a vast improvement in the economic rent from commercial fishing, individual rights-based systems have met with a great deal of resistance, including a recent congressional moratorium on the adoption of IFQs. Many fishermen adamantly oppose individual quota management, although many others express strong support (Knapp 1996, 1997). Much of the resistance to a harvest share rights system, as with earlier resistance to exclusive access rights, arises from concerns about initial allocation, fairness, employment losses, and corporate dominance.

Table 5.2 summarizes the various forms of rights-based management systems for fisheries.

Technology and the Evolution of Rights-Based Fishery Management Systems

As discussed in the introduction to this volume, property rights evolve in response to shifts in benefits and costs. This theory suggests that technology affects the emergence of rights-based fisheries management by affecting the costs of defining and enforcing rights. All else equal, technological changes that lower the costs of defining and enforcing rights should make the emergence and degree of rights-based management more likely.

Although this is a useful insight, it is insufficient to describe the role of technology in the evolution of rights-based management. Ideally, the emergence of rights-based fisheries management and the role of technology in this process should be examined within a broader view of the evolution of fisheries management institutions, systems, and regulations over time.

Fisheries management evolves within multiple institutions governing the collectively owned fishery resources. These institutions are local, regional, national, and international in scope. They include not only fisheries management agencies such as the NMFS and state departments of fish and game, but also institutions such as Congress, international treaties and commissions, and the courts. Within these institutions decisions are influenced by

Table 5.2. A Typology of Rights-Based Fishery Management Systems

Types: *Variations*	*Description*	*Examples*
Access Rights:		
Limited entry permits	Only holders of individual access rights compete in harvest.	Alaskan salmon fisheries, 1974; British Columbian salmon fisheries, 1968
Collective access rights	Only members of specific groups compete in harvest.	Washington State Indian treaty fisheries limited to tribal members; Icelandic salmon fisheries limited to landowners along rivers
Harvest Share Rights:		
Individual share quotas	Individuals hold rights to harvest shares of the annual total quota.	Most New Zealand fisheries, 1983–86; most Icelandic fisheries, 1984; U.S. surf clam fisheries, 1990; Alaskan halibut and sablefish fisheries, 1995
Collective quotas	Specific groups hold rights to harvest shares of the annual total quota; individual harvest allocation is determined by group.	Western Alaska Community Development Quotas, 1994
Cooperatives	Individuals or groups contract to convert access permits into agreed harvest shares.	Pacific Whiting Conservation Cooperative, 1998
Exclusive Rights (Fish & Shellfish):		
Leased or owned tidelands	Individuals or groups hold rights to raise and harvest within geographic areas for species with limited mobility.	Washington State oyster rearing on public and private tidelands
Public water enclosures	Individuals or groups enjoy exclusive harvest rights within artificial enclosures.	Washington State pen-reared salmon and mussels and oysters raft culture; Japanese tuna aquaculture within enclosed bays

Sources: Annala (1996); Arnason (1997); Higgins (1999); National Research Council (1999); Pearse and Wilen (1979); Singleton (1998); Young (1983).

many interest groups, including fish harvesters, fish processors, biologists, economists, elected politicians, and other groups affected by or interested in commercial fisheries. Sportfishermen, Native Americans, and environmental organizations also have interests in fisheries. Individuals within these groups are heterogeneous, with widely varying stakes in fisheries management. The multiple and interrelated objectives of the interest groups include objectives for the fishery, such as resource conservation, economic efficiency, safety of fish harvesters, allocation of economic benefits, and employment. They also include individual objectives such as fishing income, reelection, promotion, and continued agency funding. Fisheries management evolves over time in response to changes in these objectives and interests, and in response to changes in resource conditions, technology, and markets. Appreciation of the complexity of the evolution of fisheries management yields several additional insights into the relationship between technology and the emergence of rights-based systems.

First, the emergence of a rights-based system clearly cannot be explained solely in terms of potential for increased economic efficiency or net economic benefits. Fisheries management reflects many other objectives, some of which may directly conflict with efficiency, such as maintaining rural fishing employment and the use of small boats from small communities. Managers of open-access fisheries frequently craft the fishing regulations in response to the needs and demands of diverse stakeholders. The ability to respond to stakeholders gives the management agency the public influence and the political support needed to survive the budgetary and other processes that threaten small government agencies. Hence, managers are most likely to champion more exclusive harvest rights if they are seen to benefit the existing stakeholders and if the process of allocating and enforcing rights provides the management agency with a continuing supportive role in the fishery.

Second, not all technological change necessarily favors the evolution of rights-based management. Technological change may also offer fixes for problems associated with nonrights-based systems. For example, improved at-sea and onshore reporting and enforcement systems can facilitate successful conservation efforts under open-access systems. If successful management prevents the onset of a crisis in the fishery, this may delay or weaken the demand for institutional change; however, some technical innovations can weaken enforcement by helping fishing vessel operators to avoid detection by enforcement agencies. For example, shipboard radar and satellite communications may help fishing vessels learn to avoid detection and apprehension by coast guard patrol vessels.

Third, the evolution of fishery management is a dynamic process. Fishery institutions change incrementally under a given management system until a point at which pressures emerge for major change. The decisive factor in the adoption of rights-based systems may not be any change in the opportunities they offer, but rather the declining performance (shortening of seasons, increasing capitalization, declining profits) of the systems that they replace. A conservation crisis may precipitate adoption of new property rights systems because the existing stakeholders have lost the ability to support the preexisting system. For example, it is no accident that Alaska's sweeping adoption of license limitation occurred at the nadir of the salmon fishery.

Fourth, and perhaps most important, the adoption of rights-based management is not a marginal change or adjustment. It is a fundamental and profound change, with implications for the fishery and the society. The move from an open-access system to an IFQ system has the potential to change almost everything about the fishery: who fishes, when they fish, what kind of boats and gear they use, how many crew they employ and how they are paid, where they sell the fish, the prices they receive, how the fish are processed and transported to market, and who enters and profits from the fishery in the future. Because it is so fundamental a change and is so difficult to reverse, it is not entered into lightly. As a result, gradual technological change does not lead to gradual adoption of rights-based management in a fishery.

Fifth, once rights-based management has been introduced, endogenous changes in the fisheries tends to work in support of strengthening and expanding rights-based management. Rights-based management tends to erode the political influence of those who stand to lose the most from such a system. In effect, by increasing the relative importance of economic efficiency in the fishery, as reflected in part by the capitalized value of individual quotas, the dynamic of the naive theory of property rights becomes more applicable. Contributing to this process is endogenous technological development, which is driven by the need for better definition and enforcement of rights. Examples of these endogenous technological developments are improved onboard fish-weighing systems and fisherman-sponsored remote onboard fishing observation systems.

TECHNOLOGICAL INNOVATIONS THAT FACILITATE RIGHTS-BASED FISHERY MANAGEMENT

Technological innovations that facilitate the adoption of rights-based management may be grouped as enforcement technologies and utilization technologies (see table 5.3). While some new technologies may facilitate enforcement and utilization, most either improve the ability to monitor fishing activity and to enforce rules of resource use, or they improve the ability of harvesters to target and use specific species, substocks of species, or sizes of fish.

Enforcing quantitative rights in ocean fisheries can be problematic and costly, especially if the typical fishing enterprise acts opportunistically rather than cooperatively with the management program. According to Ostrom's (1990, 1995) analysis, successful governing of the commons requires that each rights holder understand the long-term benefits of collective conservation effort, and that each must believe that other rights holders are following the rules rather than concealing excessive harvests.

Table 5.3. Technologies Facilitating Rights-Based Fishery Management

Type of Technology: Examples	*Implications*	*Applications*
Enforcement:		
Real-time harvest reporting & accounting	Monitors and enforces quantitative harvest rights	Where harvests occur over long seasons with multiple deliveries (Alaskan halibut fisheries)
Remote vessel tracking	Enforces location-specific harvest rights	Where harvests cover large areas with geographically distinct stocks (Alaskan halibut fisheries)
Utilization:		
Harvest gear	Targets specific species	Where species mixing is significant (groundfish fisheries)
Harvest sampling and species identification	Monitors species-specific harvests	Where species mixing is significant (trawl gear fisheries)
Real-time weighing	Allows fishers to know how much they have harvested at any time	Where large volumes may be harvested in a short time (Alaskan halibut fisheries)

In the case of individual quotas, following the rules means harvesting within the allotted share of the annual aggregate catch quota. The typical harvester commits to following the rule so long as other harvesters display credible commitments to also follow the rule. This credible commitment is tied to monitoring and enforcement actions, including graduated sanctions such as increasingly stiff fines for repeated offenses of fishing regulations. For example, a fisherman who cannot accurately determine whether he is on a particular fishing ground and believes others also cannot determine such facts is unlikely to commit to rules dealing with exclusive rights or other regulations pertaining to such fishing grounds. As a result the emergence of rights-based management is facilitated by technological improvements that can create credible monitoring and enforcement systems that encourage voluntary conservation rule–following behavior.

Utilization technologies facilitate the precision with which fishermen are able to apply specific harvest rights to different species and thus affect the extent to which more comprehensive and specific harvest rights can be defined. These technologies provide economically efficient fisheries conservation objectives for multiple species. In particular, they affect the extent to which rights-based systems can avoid the problem of bycatch, defined as the inadvertent capture of nontarget species. Improved navigation, gear design, and fish handling ability can reduce bycatch and protect individual harvest rights from bycatch.

For example, assignment of north Pacific halibut fishing rights to vessel owners using longline gear requires that groundfish trawlers be excluded from halibut fishing. This requires that trawlers be able to capture fish allocated to them without taking halibut. This is limited by the degree to which the trawl gear encounters, captures, and kills halibut; the ability of trawler crews to discard halibut without injury; and the ability of the trawl operator to avoid times and places where halibut mix with other groundfish species.

If harvest rights are created only for target species, managers may still need to restrict harvesters' flexibility in when and how they may fish for target species as a control on bycatch. Copes (1986, 288–89) suggested that "what really counts in rationalizing the fisheries is not what property rights have been installed, but what externalities remain or are newly created by the particular form of partial property rights introduced." In theory, the bycatch problem could be addressed by creating harvest rights for both the target species and the bycatch species, providing fishermen the incentive to target fish of each species in proportion to their individual catch rights. The effectiveness of this approach is limited by the extent to which fishermen are able to match their catches of different species to their quota holdings. Separation of species into exclusive harvest rights would be improved by technological ability of fishermen to target species more accurately.

IMPROVED HARVEST REPORTING

Effective enforcement of quantitative harvest rights requires a method by which enforcement personnel can determine whether a fisherman has exceeded his quota. The technical complexity of enforcement depends upon characteristics of the fish and the fishery. The first individual quota program to be approved under the Magnuson-Stevens Act, the surf clam/ocean quahog fishery off the Mid-Atlantic states, addressed the harvest reporting problem in a simple, low-tech way. Clam quotas are defined in bushels that are aggregated and transported in larger units called cages. Each year quota holders receive single-use cage tags for the amount of their quota. These tags must be affixed to cages when clams are harvested. Any cage with a valid tag is deemed to contain legal clams. Vessel owners and dealers are also required to report clams landed and purchased. Enforcement is simplified by the fact that there were fewer than 140 vessels harvesting over a relatively small geographic range (National Research Council 1999).

The halibut fishery off Alaska, with annual harvests valued at between $50 and $100 million, is the most valuable American fishery for which share-rights management has been adopted. Since World War II, the Pacific halibut fishery had been regulated by an annual harvest quota, which is taken by longline gear (hook and line).[2] During the two decades before the IFQ program, increasing prices and fairly stable annual harvest quotas encouraged the competitive, open-access fishing fleet to grow from roughly 1,000 vessels in 1975 to more than 3,500 vessels in 1993. To prevent overfishing, the fishing seasons were steadily shortened, from about 125 days in 1975 to just 2 or 3 days by the late 1980s (Berman and Leask 1994). Investments in vessels and fishing gear grew with halibut prices, increasing total costs, and net fishing incomes remained below their potential. In 1995 the North Pacific Fishery Management Council implemented an IFQ allocation plan that assigned individual annual quota shares to more than 4,000 individuals based upon documented harvest shares from 1984 to 1990. The overall halibut quota is divided among four vessel categories, based on length and freezing capacity, and eight geographic areas. Among the initial allocations, more than 3,000 yielded harvest rights for less than 1,000 pounds, and 69 were for more than 50,000 pounds (NMFS 1995b).

Although the IFQ program offered the promise of reduced costs, higher market value, and a safer fishery, all of which were subsequently realized, it posed a significant enforcement challenge. Prior to the IFQ program, except for during and immediately after the short halibut seasons, any vessels with halibut onboard were in clear violation of management regulations. In contrast, under the IFQ program, during the eight months between March 15 and November 15 when IFQ harvests were permitted, vessels with halibut onboard might be fishing either legally or illegally, depending on whether those onboard held remaining unfished individual quota. Credible implementation of the IFQ program depends on the ability of enforcement agents to monitor quota holdings and catches in real time. It fell to the newly established Restricted Access Management (RAM) Division at NMFS's regional office in Juneau, Alaska, to devise such a system.

To address this challenge, RAM called for joint reporting of halibut landings by fishermen and registered buyers, which includes fish processors, distributors, restaurants, and fishermen selling directly to consumers. Landing reports are submitted electronically using automated transaction terminals (similar to automated teller machines [ATMs]) connected via telephone modems set to dial toll-free numbers to RAM computers in Juneau, Alaska. The NMFS originally purchased 450 ATM units and issued them free of charge to registered buyers. Each year fishermen are issued plastic cards with magnetic identification strips. Landing reports are made electronically using transaction terminals and personal cards with a four-digit personal identification number for security, similar to the use of a credit card or an ATM transaction. When the landing report is entered, a receipt is printed in three copies and the IFQ holder keeps a copy. If there is a problem with the transaction, such as the fisherman exceeding his IFQ, the registered buyer will be advised to contact NMFS enforcement before proceeding. The transaction terminal can also be used to report the remaining annual balance of IFQ pounds for both the IFQ account and the vessel (NMFS 1995a).[3]

Although there were initial difficulties with electronic reporting that were attributable to the sometimes primitive Alaska communications infrastructure, these problems were largely overcome. Electronic reports comprised an average of 53 percent of IFQ transactions in 1996, 89 percent in 1997, and 98 percent in 1998.[4]

The electronic reporting system not only has provided a method by which enforcement officials can check the unharvested quotas of vessels, but also proved to be adaptable for other enforcement purposes. For example, the system allows for monitoring and immediate detection and auditing of potentially suspicious landing reports, such as inconsistencies between ports of landing and reported harvest areas or low reported landings that happen to conform too exactly to remaining IFQs.

The electronic reporting for the Alaska halibut IFQ program provides an example of the adaptation of exogenous technology to the needs of a rights-based management system. It would be difficult to argue that the existence of this technology was a necessary condition for the implementation of the program; a more cumbersome logbook recording system might have worked adequately. But the demonstration of the potential for real-time reporting and successful enforcement in a widely dispersed fishery with thousands of participants may increase the likelihood that similar programs will be adopted in other fisheries. In addition, the implementation of the Alaska halibut IFQ program has led to the endogenous development of technology to advance electronic reporting. For example, personal computers and the Internet are being used more frequently.

Monitoring and Enforcement via Satellite

The predominant method of surveillance and enforcement in the coastal oceans and EEZ has historically involved visual or radar detection of fishing vessels from patrol ships and airplanes. The high cost of such surveillance limits the degree to which the federal government can effectively detect and intercept foreign or domestic vessel fishing in violation of international or domestic law. Recently, a number of technical innovations have become available to fisheries enforcement personnel that decrease the costs of maintaining widespread and near-continuous surveillance of the EEZ. This should increase the degree of exclusivity of the nation's collective property right in ocean fish stocks within the EEZ.

The National Oceanic and Atmospheric Administration's (NOAA) enforcement arm in Hawaii has a particularly difficult task in monitoring the huge EEZ around the Hawaiian Islands. In recent years they have been inducing foreign fishing vessels seized in illegal fishing to voluntarily enroll in a high-tech vessel management system (VMS). The VMS involves a land-based receiving station connected via satellite to transponders (mobile transceiver units) aboard vessels at sea. The transponders in turn receive location data from global positioning system (GPS) satellites and answer interrogations from the land station with the vessel's identity, position, speed, and course. This allows agents at the receiving station to monitor in near real time a vessel's path. Further, because the details of ship movements during longline fishing operations for tuna and billfish have a characteristic signature, the agents can determine with fair accuracy when the vessels are fishing. If unauthorized fishing is detected in the EEZ, an airplane will be dispatched to photograph and verify the illegal operation. This is feasible because it takes hours to deploy and retrieve a commercial longline set. Federal enforcement agents hope to establish the signature analysis alone as prima facie evidence of a violation. This reduces the cost of enforcement, but in the case of foreign fishing vessels, does not overcome the problem of jurisdiction to prosecute them. The jurisdictional issues are subject to ongoing negotiations.[5]

Although only a subset of foreign fishing vessels currently carries the satellite transponders, this technology could facilitate the broader and more effective enforcement of the EEZ, much as shore cannons enabled European nations to claim three-mile zones in the seventeenth century. This method is particularly important in portions of the EEZ that are distant from land, and therefore expensive to monitor by airplane or ship, but which contain high-value marine fish concentrations. The EEZ around Hawaii is one such area, and the central Bering Sea, especially along the maritime boundary between the United States and Russia, is another location where such a system could prove effective.

The same type of satellite positioning and identification system could be useful in domestic fisheries, for instance, the Bering Sea groundfish and crab fisheries where managers open and close specific areas during the fishing season.

Weighing and Sampling Technology

A formal IFQ system or a cooperative system of allocated quotas requires a reliable system of measuring and reporting actual catches in order to maintain credibility among fishermen and to encourage cooperative compliance with fishing rules. Where multiple species are caught, the requirement is to measure total volume and species composition of the catch. Further, when some species or size categories of fish are protected from harvest and must be returned to the sea, the incidence of bycatch and its disposition need to be documented.

Batch sampling and other approximation methods are often used to establish the weight and species composition of mixed loads of fish, especially on large groundfish trawlers. This means of measuring fish caught at sea is often only roughly accurate. Conventional weight scales prove inaccurate when used on rolling and pitching ships at sea. In recent years there have been several technical improvements. Strain gauges on the winches can help with total weight measurement, and motion-compensated flow scales or motion-compensated flatbed scales can provide accurate weight measurement of fish as they are moved from the fish deck to an onboard processing plant or storage holds. In multispecies fisheries, however, accurate measurement of total weight is insufficient to monitor and enforce rights to an individual species catch. When large volumes of mixed species of fish are unloaded from nets during fishing operations, detecting species composition is difficult. Sampling of portions of the catch by observers provides a statistical estimate of species composition.

For example, in 1991 the North Pacific Fishery Management Council adopted the Vessel Incentive Program, which was intended to increase the utilization of groundfish harvest allowances while keeping the bycatch of Pacific halibut and red king crab within strict limits. Each vessel carried an observer to estimate bycatch rates and report by a satellite communication system on a weekly basis. Vessels that exceed the standard bycatch rate were to be penalized. Renko (1998) concluded, however, that the system failed to affect vessel operations or bycatch rates because vessel operators were able to conceal bycatches from observers, and where at-sea samples indicated high bycatch rates, the statistical information provided by partial samples was considered too weak to stand up in court. As a result, very few of the suspected violators of the bycatch standards have been prosecuted, and the program creates little or no economic incentive to the vessel operators.

Thinking of the Future

We can envision vast technological improvements that could expand the potential for rights-based fisheries management in the future: improvements in technologies for monitoring ship position and fishing activity, measurement of catch and species composition, and real-time quota adjustments for in-season management of salmon.

Satellites currently used for spying on foreign military facilities could greatly improve the ability to monitor ship position and fishing activity. These satellites can provide very accurate images down to one-meter scale. Such accurate satellite images of fishing vessels could provide information concerning types of fishing gear used and, when fish are on deck, some information about species being caught. Simple random monitoring could provide authorities with enough information to prosecute violators of closed seasons and fishing gear regulations. The very existence of such information-gathering capability would be a deterrent to cheating on negotiated fishing rules. Currently these satellites are not available for fishery enforcement duty.

Video cameras could be adapted to measure and monitor fish catch and species composition. On larger vessels this is currently carried out by observers, as required by management regulations, but space limitations prevent housing official observers aboard smaller fishing vessels. Currently, vessels less than 65 feet long are not required to carry observers in the fisheries off Alaska, whereas vessels longer than 125 feet always carry observers, and many large catching and processing vessels carry two observers. Lack of observer coverage on the numerous smaller vessels reduces accurate measurement of mixed catches, bycatch, and fishing gear use.

To address this problem, Canada's Pacific Blackcod Fisherman's Association, in cooperation with Archipelago Marine Research Ltd., developed a video-surveillance system to replace at-sea observers (McElderry et al. 1999, 63). The system consists of a battery/backup power source, which ensures that the system will operate if the vessel's power source is cut; a combination satellite positioning system/video recorder system that monitors the vessel position on a video screen; and a camera placed to view the fishing deck. The camera is programmed to take a picture every ten seconds. The videotape is monitored after each trip to ensure that the vessel was fishing in an authorized area and that it is not off-loading fish at sea. The video does not currently have adequate resolution to identify species or estimate catch weight. Each vessel owner can rent one of these units from the fisherman's association for use in an experimental seamount black cod fishery. The incentive to do so is that this is a low-cost (about C$1,500 for a thirty-day trip) alternative to carrying a mandatory observer (about C$8,000).

Increased resolution of the video images raises other exciting possibilities. For example, the estimation of species composition of catches might be accomplished without labor-intensive observer coverage. Improved optical pattern recognition software combined with digital cameras could be used to estimate the species in mixed harvests aboard fishing vessels. This would facilitate accounting for catches and monitoring of bycatch.

Until recently, economists and fisheries managers have assumed that individual quota management could not be adapted to fisheries requiring real-time frequent adjustment of fishing intensity. For example, North Pacific salmon fisheries are managed to allow the fish that escape the nets of commercial fishermen to proceed upriver to their spawning grounds. Depending upon the actual size of the run, managers adjust the length of fishing seasons to allow fishermen to catch surplus fish after escapement goals have been met. Copes (1986, 283) has argued that "individual quota management is inherently unsuited to fisheries where the catch is residual to a managed escapement target."

With radio and real-time electronic reporting technology, there is no longer any reason for individual fishing quotas to be calculated seasonally. It is possible to envision a salmon fishery in which managers calculate a total daily quota to be announced by radio, with individual daily quotas calculated as shares of the total daily quota. These total daily quotas might on some days be very small, in which case quota holders could arrange through preseason contracting to save on fishing expenses by selling their quotas to a few designated vessels. On other days, managers might allow for unlimited fishing to utilize the full extent of fishing capacity. A daily fishing quota system has been proposed for Alaska's valuable but overcapitalized Bristol Bay salmon fishery, but this proposal has gained little support so far (Vander Lind 1999).

CONCLUSION

We anticipate that rights-based fishery management will continue to develop as existing systems demonstrate the potential for economic benefit, political and legal challenges are overcome, and technological developments facilitate monitoring, enforcement, trading in rights, and the ability to target specific species. We have shown that institutions creating more exclusive harvest rights systems have used a range of technological developments to facilitate monitoring and enforcing individual quotas. This is demonstrated by the examples of IFQ programs for halibut and surf clams. In contrast, the fisheries cooperatives for Pacific whiting and Alaska pollock are examples of private contracting, within a permissive legal regime, for more exclusive rights that do not rely on significant new technology or government regulation.

New technology can improve the economic benefits from individual harvest rights, but it plays a limited role in the political decision to adopt rights-based approaches to fish conservation. The collective decision processes are at least as concerned about distributional consequences as they are about aggregate economic benefits, and technology seems to offer little help in resolving these issues. Concerns have been strongly expressed for the equity of initial quota distribution and for redistribution of resource rents due to industrial restructuring under IFQs (Matulich, Mittelhammer, and Reberte 1996; National Research Council 1999, 4). These concerns were apparently responsible for the congressional moratorium (Sustainable Fisheries Act of 1999) on additional IFQ systems in U.S. fisheries. That distributional problems would jeopardize creation of more exclusive property rights in common pool resources was predicted by Libecap (1989), Ostrom (1990), and others. Thus the development of more exclusive property rights in fisheries may require institutional means of assuring major stakeholders in the fisheries that their economic and social positions will not be reduced by the property rights system. One such means would be to create a legal plan in which the competing claims can be negotiated to a consensus allocation. This is essentially what happened in the privately negotiated Pacific Whiting

Conservation Cooperative. However the distributional issues are resolved, the successful development of rights-based systems depends on mutually ensured compliance with harvest entitlements based, at least partly, on credible monitoring and enforcement mechanisms.

In the broad sweep of property rules and rights regimes, the technical facilitation of enforcement and utilization plays a subsidiary role to the demonstration of significant gain from rights creation. In Alaska, for example, the timing and content of the halibut IFQ program was most certainly influenced crucially by observed economic performance of British Columbia's sablefish and halibut fleet. Under the Canadian individual vessel quota program, which was established in 1990, Canadians were harvesting the same species of fish as Alaskans while receiving significantly higher prices and enjoying much greater flexibility in fishing operations. The contrast in economic benefits of fishing between Alaska and British Columbia must have been an encouraging factor in the adoption of IFQs in Alaska. Still, the technical ability to monitor landings on a near real-time basis and to track fishing operations at sea may be crucial to credible enforcement of individual harvest rights, especially in fisheries like halibut that are conducted over wide-ranging fishing grounds.

NOTES

1. There is ample evidence of previous local customs that performed essentially as limited access systems. Ancient rules of capture in places like the Columbia River salmon fishery clearly limited access to choice harvesting sites (Higgs 1982). Many such local systems were destroyed with the expansion of Western European legal and cultural traditions.
2. Halibut is under the jurisdiction of the International Pacific Halibut Commission, created in 1923 by a joint treaty between the United States and Canada. The commission establishes gear restrictions and area-specific fishing seasons and annual total allowable catches. The U.S. government, through the North Pacific Fishery Management Council, establishes additional regulations to meet the total allowable catches and to allocate harvests in U.S. waters. See Knapp (1996) for a description of the Alaska halibut IFQ program. A similar program was also adopted for the Alaska sablefish (black cod) fishery.
3. To facilitate selective monitoring of landings, fishermen were also required to notify NMFS at least six hours prior to making a landing, and to begin landings between 6:00 A.M. and 6:00 P.M. Landings were not restricted to specific ports, however, as had been done in the halibut IFQ program implemented in British Columbia in 1991.
4. Jessica Gharrett, data manager, NMFS Restricted Access Management Division, Juneau, Alaska, memorandum to Phil Smith, RAM program administrator, "Request for Transaction Terminal Status," June 1, 1999.
5. Paul Ortiz, enforcement attorney, U.S. Department of Commerce, NOAA, Southwest Region, Long Beach, California, e-mail communication, October 18, 1999.

REFERENCES

Annala, John. 1996. New Zealand's ITQ System: Have the First Eight Years Been a Success or a Failure? *Reviews in Fish Biology and Fisheries* 6(1): 43–62.

Arnason, Ragnar. 1997. The Icelandic Individual Transferable Quota System: Motivation, Structure, and Performance. In *Global Trends: Fisheries Management*, ed. Ellen K. Pikitch, Daniel D. Huppert, and Michael P. Sissenwine. Bethesda, Md.: American Fisheries Society, 225–36.

Berman, Matthew, and Linda Leask. 1994. On the Eve of IFQs: Fishing for Alaska's Halibut and Sablefish. *Alaska Review of Social and Economic Conditions* 29(2): 1–20.

Boyd, Rick O., and Christopher M. Dewees. 1992. Putting Theory into Practice: Individual Transferable Quotas in New Zealand's Fisheries. *Society & Natural Resources* 5(2): 179–98.

Burke, William. 1994. *The New International Law of Fisheries: UNCLOS 1982 and Beyond.* Oxford, England: Clarendon Press.

Casey, Keith E., Christopher M. Dewees, Bruce R. Turris, and James E. Wilen. 1995. The Effects of Individual Vessel Quotas in the British Columbia Halibut Fishery. *Marine Resource Economics* 10(3): 211–30.

Christy, Francis T. 1997. Economic Waste in Fisheries: Impediments to Change and Conditions for Improvement. In *Global Trends: Fisheries Management*, ed. Ellen K. Pikitch, Daniel D. Huppert, and Michael P. Sissenwine. Bethesda, Md.: American Fisheries Society, 28–39.

Copes, Parzival. 1986. A Critical Review of the Individual Quota as a Device in Fisheries Management. *Land Economics* 62(3): 278–91.

Crutchfield, James A., and Giulio Pontecorvo. 1969. *The Pacific Salmon Fisheries: A Study of Irrational Conservation.* Baltimore: Johns Hopkins University Press.

Food and Agricultural Organization of the United Nations (FAO). 1998. *The State of World Fisheries and Aquaculture.* Rome: FAO Fisheries Department.

Gardner, Michael. 1989. The Enterprise Allocation System in the Offshore Groundfish Sector in Atlantic Canada. In *Rights Based Fishing*, ed. P. A. Nehrer, R. Arnason, and N. Mollett. NATO ASI Series. Dordrecht, Netherlands: Kluwer.

Geen, Gary, and Mark Nayar. 1989. Individual Transferable Quotas in the Southern Bluefin Tuna Fishery: An Economic Appraisal. In *Rights Based Fishing*, ed. P. A. Neher, R. Arnason, and N. Mollett. NATO ASI Series. Dordrecht, Netherlands: Kluwer Academic Publishers, 355–81.

Gordon, H. Scott. 1954. The Economic Theory of a Common Property Resource. *Journal of Political Economy* 61: 124–42.

Higgs, Robert. 1982. Legally Induced Technical Regress in the Washington Salmon Fishery. *Research in Economic History* 7: 55–86.

Higgins, Kollin. 1999. The Pacific Whiting Conservation Cooperative: As Good as IFQs? Master's thesis. School of Marine Affairs, University of Washington, Seattle.

Homans, Francis R., and James E. Wilen. 1997. A Model of Regulated Open Access Resource Use. *Journal of Environmental Economics and Management* 32: 1–21.

Huppert, Daniel D., Gregory M. Ellis, and Ben Noble. 1996. Do Permit Prices Reflect the Discounted Value of Fishing? Evidence from Alaska's Commercial Salmon Fisheries. *Canadian Journal of Fisheries and Aquatic Sciences* 43(4): 761–68.

Knapp, Gunnar. 1996. Thalassorama Alaska Halibut Captains' Attitudes Towards IFQs. *Marine Resource Economics* 11(1): 43–55.

———. 1997. Thalassorama: Initial Effects of the Alaska Halibut IFQ Program: Survey Comments of Alaska Fishermen. *Marine Resource Economics* 12(3): 239–48.

Libecap, Gary D. 1989. *Contracting for Property Rights.* Cambridge, England: Cambridge University Press.

Matulich, Scott C., Ronald C. Mittelhammer, and Carlos Reberte. 1996. Toward a More Complete Model of Individual Transferable Fishing Quotas: Implications of Incorporating the Processing Sector. *Journal of Environmental Economics and Management* 31: 112–28.

McElderry, Howard, William A. Karp, John Twomey, Mandy Merklein, Vicki Cornish, and Mark Saunders. 1999. Proceeding of the First Biennial Canada/U.S. Observer Program Workshop. *NOAA Technical Memorandum* NMFS-AFSC-101. Seattle: U.S. Department of Commerce, National Oceanic and Atmospheric Administration, National Marine Fisheries Service, Alaska Fisheries Science Center.

Moloney, David G., and Peter H. Pearse. 1979. Quantitative Rights as an Instrument for Regulating Commercial Fisheries. *Journal of the Fisheries Research Board of Canada* 36: 859–66.

National Marine Fisheries Service (NMFS). 1995a. *The IFQ Program: Under Way.* Informational pamphlet. Juneau: Alaska Regional Office, Restricted Access Management Division.

———. 1995b. Quota Share Holder File as of November 22, 1995. Juneau: Alaska Regional Office, Restricted Access Management Division.

———. 1999. *Fisheries of the United States, 1998.* Washington, D.C.: National Oceanic and Atmospheric Administration.

National Research Council. 1999. *Sharing the Fish: Towards a National Policy on Individual Fishing Quotas.* Washington, D.C.: National Academy Press.

O'Connell, Daniel P. 1982. *The International Law of the Sea,* ed. I. A. Shearer. New York: Clarendon Press.

Ostrom, Elinor. 1990. *Governing the Commons: The Evolution of Institutions for Collective Action.* Cambridge, England: Cambridge University Press.

———. 1995. Designing Complexity to Govern Complexity. In *Property Rights and the Environment,* ed. Susan Hanna and Mohan Munasinghe. Washington, D.C.: World Bank, 33–45.

Pearse, Peter H., and James E. Wilen. 1979. Impact of Canada's Pacific Salmon Fleet Control Program. *Journal of the Fisheries Research Board of Canada* 36(7): 764–69.

Renko, Becky L. 1998. *Management of Pacific Halibut Bycatch in the Alaskan Groundfish Fishery 1990–1996: The Vessel Incentive Program.* Master's thesis, School of Marine Affairs, University of Washington, Seattle.

Scott, Anthony T. 1955. The Fishery: The Objectives of Sole Ownership. *Journal of Political Economy* 63: 116–24.

Singleton, Sara. 1998. *Constructing Cooperation: The Evolution of Institutions of Co-management.* Ann Arbor: University of Michigan Press.

Vander Lind, Larry. 1999. Quota Share Buyback Program for Harvesting Salmon by Driftnetters in Bristol Bay. Proposal distributed at Alaska Salmon Forum '99, Anchorage, March 12.

Young, Oran R. 1983. Fishing by Permit: Restricted Common Property in Practice. *Ocean Development and International Law* 13(2): 121–70.

6

The Potential of High Technology for Establishing Tradable Rights to Whales

Gregory B. Christainsen and Brian C. Gothberg

We do not own this Earth, we simply borrow it from our children. . . .
Am I prepared to die to defend the whales? The answer is yes.
I would consider it an honour to die in their defense.[1]
—Captain Paul Watson
Founding director, Greenpeace
Founder, Sea Shepherd Conservation Society

No cry of environmentalists has been as loud as "Save the Whales!" Indeed, several stocks of whales were decimated in the early and middle parts of the twentieth century. Under the circumstances, it was quite plausible to regulate whaling. In what follows, it will be seen that certain advances in technology played a major role in depleting whale stocks, but that other technological developments may make possible some deregulation of whaling in the future.

PROPERTY RIGHTS AND REGULATION FOR WILDLIFE

To help organize the discussion, we will first outline a theory of property rights and regulation for wildlife. Then we will note how the theory applies to whales. Following Anderson (1998) and Lueck (1991), the optimal regime of property rights and/or regulation for a creature depends on the pattern of land (or water) ownership and the creature's territorial needs. If the only use for land were the provision of habitat for the creature, it would be efficient to define landownership rights that corresponded to the animal's requirements. A single landowner could own an entire stock of wildlife, but if there are many possible uses for land, the ideal plot size may be smaller than the wildlife's requirements. The wildlife might then range over an area that exceeds the domain of any single landowner.

An organized group of landowners might still play a role in governing access to the creature in such cases, but much depends on the transaction costs involved. At one extreme, it might be possible at relatively low cost to define and enforce property rights to individual members of the stock of wildlife at issue. If, at the same time, contracts between the wildlife owners and landowners can be negotiated and enforced without much difficulty, the wildlife regime need not involve much public regulation. If ownership of individual animals is feasible, but contracting with landowners is costly, it would be possible under the principle of eminent domain to give the creature the right of way to the territory in question.[2]

Even if it is not possible to have ownership of individual animals, landowners may be able at low cost to contract with one another to form an organization to sell hunting or viewing rights to animals (e.g., elk) that stray onto their collective property. Going one step further, a broker can act as an intermediary between wildlife consumers and a group of landowners. At the other extreme, the costs of domesticating animals and of contracting among multiple landowners might be so high that they exceed the value of the wildlife. In such cases, the wildlife might be left to exist as an open-access commons. Open access, however, can result in the wildlife stocks being depleted or even extinguished, so public regulation of hunting becomes appealing as a way to prevent the wildlife's value from being dissipated.

Tradable harvesting quotas are attractive in the event that some hunting can be safely allowed, but establishing property rights to the stock itself is not feasible. Tradable quotas put a limit on the total amount of hunting, and they encourage high-cost, low-profit harvesters to sell their entitlements to lower-cost, higher-profit producers or to environmental groups who would retire the permits (as in Anderson and Leal 1997, 103).

Keeping these considerations in mind, we can appreciate the difficulties historically encountered with respect to whales. On the one hand, their territorial requirements are typically much greater in size than the water holdings of any person, private organization, or association of private organizations. On the other hand, it has rarely been possible to establish property rights to individual animals except after they have been harpooned. Whales therefore existed for many years as an open-access commons. Pressures to regulate whaling emerged, however, as some stocks of whales became severely depleted. Since the territorial requirements of whales typically exceed a single country's jurisdiction, regulation became international in nature. The regulation has not had the unanimous approval of the nations involved, and the controversies that it has provoked have led some countries to secede from the regulatory management.

Early Commercial Whaling: The Impact of Technology under Open-Access Conditions

Commercial whaling became significant in the 1600s with the establishment of large whaling fleets by the English and the Dutch. As nearby whale populations began to be depleted, vessels ventured out into the Atlantic as far as Greenland (Jackson 1978).

In the 1700s, vessels from the American colonies also became common in the Atlantic. The United States in due course dominated the industry. By the mid-1800s, there were more than 700 American ships involved, with most of them sailing out of New England ports such as New Bedford and Nantucket. The technology for capture was primitive, and whales were harpooned from open boats (Starbuck 1964).

The industry changed in 1864 with the invention of a whaling cannon by a Norwegian, Svend Foyn. This cannon was capable of launching harpoons with an explosive charge. Prior to this, whalers had to focus on slower-moving species of whales such as the bowhead, sperm, gray, and humpback so that kills could be made with thrown harpoons. Foyn's innovation, along with the subsequent evolution of faster, motor-powered boats, enabled whalers to go after swifter species such as the blue, sei, and fin whales (Tonnessen and Johnsen 1982).

With the advent of underwater sonar, helicopters, and weaponry of increasing firepower, twentieth-century whalers were able to harvest tremendous numbers of their prey. Compressors allowed whalers to pump pressurized air into harpooned whales that otherwise would have sunk, thereby making the recovery and processing of the rorquals much easier. Perhaps the decisive innovation was the invention in 1925 of the stern slipway for factory ships (Tonnessen and Johnsen 1982). This enabled whalers to haul animals on deck and process them without a land-based whaling station. Ships now flourished in the whale-rich pelagic areas of the Antarctic. By the 1930–31 whaling season, there were forty-one factory ships in operation, and they harvested more than 37,000 whales. A blue whale could subsequently be dismembered, butchered, and pressure-cooked to produce meals and oils in forty-five minutes (*New Encyclopaedia Britannica* 1991).

THE EMERGENCE OF REGULATION

As technological advances reduced the costs of whaling, whale populations dwindled and political discussions multiplied. Beginning in the 1920s, members of the whaling industry (more so than environmentalists) sought a formal mechanism to conserve whale stocks. The effort was made primarily out of considerations of self-interest, not quasi-religious concerns about animal rights. Indeed, the discussions were driven by two economic considerations: the desire to form a cartel to support the price of whale oil and the necessity of preventing overexploitation of the whales.

In 1931 representatives from several whaling nations met in Geneva under the auspices of the League of Nations and agreed to terminate the harvesting of right whales, so-called because their ample size and slow swimming speed made them the right whale to catch. An exception was made for harvests by indigenous peoples who wanted to hunt whales primarily to support subsistence consumption (Gambell 1993).

Toward the end of World War II, discussions began again in earnest and climaxed in Washington, D.C., with the 1946 International Convention for the Regulation of Whaling (ICRW) (U.S. Department of State 1946). The parties to the ICRW also agreed to establish the International Whaling Commission (IWC), which began with fourteen member-countries. The IWC was empowered to regulate the whaling industry, but was not given any authority to amend the ICRW itself.

The IWC originally established a whaling limit for large cetaceans (e.g., those with members that are typically more than four meters [thirteen feet] long) through what were called blue whale units (BWUs). In terms of yielding whale oil, it was estimated that on average one blue whale is equal to 2 fin whales, 2.5 humpbacks, or 6 seis. A catch limit of 16,000 BWUs, which was estimated to be 46 percent of the harvest during the 1930–31 season, had already been agreed upon by the participating nations by 1945 (Tonnessen and Johnsen 1982).

This limit was, unfortunately, inconsistent with the conservation of whale stocks. Stocks continued to shrink. Falsification of catch data by the Soviet whaling fleet was another problem. To cite the most egregious deception, the Soviets reported the killing of 2,700 humpback whales in the Southern Hemisphere from the late 1940s to the early 1960s. The actual number hunted was closer to 48,000 (Clapham 1997, 114). Finally, in 1963, humpbacks were given complete protection south of the equator. Blue whales were largely protected in the area south of 40° south latitude. Complete protection for blue whales in the Southern Hemisphere was approved in 1967. In 1972, an International Observer Scheme was implemented whereby IWC inspectors were stationed at the whaling operations of member-states to confirm compliance with regulations.

FROM REGULATION TO A MORATORIUM

At the 1972 and 1973 annual meetings, the United States called for a moratorium on all commercial whaling. This proposal was voted down on each occasion, whereupon a growing number of environmental activists threatened to boycott goods from whaling nations (e.g., Norway, Iceland, Japan, and the Soviet Union). In 1974 Australia proposed that commercial whaling be stopped only in the case of whale stocks that had been badly depleted. Under this proposal, which was endorsed by the United States, whaling could continue with respect to other stocks of whales as long as it were done on a sustainable basis. The Scientific Committee of the IWC accepted the idea, and a so-called New Management Procedure (NMP) was implemented during 1975 and 1976 (Aron, Burke, and Freeman 1999).

Under the NMP, scientific data had to be compiled on a regular basis regarding the sizes of the various whale stocks and their recent rate of growth. It is understood that the sizes of the various stocks cannot be estimated with precision. But most members of the Scientific Committee did not believe that the uncertainty surrounding the estimates was great enough to warrant a halt to all whaling.

Antiwhaling forces were nevertheless able to gather strength. These forces were fueled not only by allegations of serious data problems but also by an emerging animal rights movement and a commission-packing scheme. Animal rights activists led by Greenpeace packed the IWC by recruiting additional nonwhaling countries to become members (Spencer and Bollwerk 1991). The enlarged IWC then approved a moratorium on all commercial whaling in 1982, which became effective in 1986.

Canada seceded from the commission when the moratorium was passed; Iceland withdrew in 1992. In 1999 its parliament resolved that the country should resume commercial whaling. Norway registered a formal objection to the moratorium early on and resumed commercial whaling in 1993. In 1987, Japan submitted a proposal to kill 900 whales annually for research purposes. The IWC passed a nonbinding resolution to condemn the proposal, whereupon Japan countered with a proposal for a harvest of 300 whales. The IWC condemned the smaller harvest as well, but Japan decided to go ahead with it. A harvest has taken place every year since then.

Antiwhaling countries agreed to reconsider the moratorium in return for updated and comprehensive estimates of whale stocks. The Scientific Committee of the IWC was also asked to devise an improved management process, a gesture that suggested a possible resumption of commercial whaling if stocks seemed to be abundant. Under the Revised Management Procedure (RMP), which obtained the unanimous approval of the Scientific Committee in 1993, the objective would be to maintain each whale stock at between 54 and 72 percent of its historical level. A stock of about 60 percent of this level is typically estimated to produce the maximum sustainable yield of whales.

Antiwhalers and whalers alike have usually regarded the RMP's approach as conservative (i.e., risk averse with respect to the conservation of whales) (Motluk 1996, 12), but the IWC resisted the actual implementation of the

RMP. Instead, in 1994, without the approval of the Scientific Committee, it created a sanctuary extending south from 40° south latitude toward Antarctica. No commercial whaling is permitted in this vast area despite the fact that the Scientific Committee estimated (through the use of studies financed by the Japanese) that the area probably contains more than 750,000 minke whales. Philip Hammond, the chairman of the Scientific Committee, resigned because of the commission's unwillingness to implement the RMP.

The IWC continues to resist any resumption of commercial whaling. The United Kingdom in particular has been quite candid in admitting that its support of the IWC's policy no longer has any basis in science (Motluk 1996, 12). A U.S. whaling commissioner made a similar admission in 1991 (Aron, Burke, and Freeman 1999, 24). There is simply a view that killing whales is wrong, with exceptions made for hunting by aboriginal peoples and for certain research purposes.

Recent years have seen the development of the Revised Management Scheme (RMS) (IWC 1999), which would supplement the RMP with sophisticated enforcement measures such as satellite monitoring of whaling vessels, onboard international observers, and most probably (it is still under discussion) DNA testing of whales and whale products.[3] Again, however, the IWC has taken no steps to make the RMS a reality, with some commentators charging disingenuousness; member-countries of the IWC can appear to be flexible by going forward with the development of measures like the RMP and the RMS, but they actually have no intention of giving their approval to any commercial whaling (Aron, Burke, and Freeman 1999).

WHALE POPULATIONS AND U.S. LAW

In 1989 the IWC decided not to provide whale population statistics except for species/stocks that have been assessed in detail and for which there is statistical certainty. The available estimates are shown in table 6.1.

Under customary international law, each country has control over the whales within its jurisdiction, which now extends 200 miles from its coastline. Whaling is permitted on the high seas unless the country of registry (i.e., the flag under which a whaling vessel is flying) has specified otherwise.

The United States is one of the countries that, for the most part, now forbids whaling. The best known American legislation to protect animal and plant life is the Endangered Species Act, passed in 1973; the Marine Mammal Protection Act (MMPA) was passed during the previous year. The MMPA established a moratorium on the taking and importing of marine mammals, their parts, and products, even if the species in question is not endangered. Eastern Pacific gray whales, for example, are no longer deemed endangered by the U.S. government, but it is still generally illegal to hunt them. Violations of the MMPA may result in fines of up to $100,000 and one year's imprison-

Table 6.1. Estimates of Whale Populations

Species *Location*	*Time Period*	*Population*	*95% Confidence Interval*
Minke whales:			
Southern Hemisphere	1982–1989	761,000	510,000–1,140,000
North Atlantic*	1987–1995	149,000	120,000–152,000
North Pacific	1989–1990	25,000	12,800–48,600
Fin whales:			
North Atlantic	1969–1989	47,300	27,700–82,000
Gray whales:			
Eastern North Pacific	1987–1988	21,000	19,800–22,500
Bowheads:			
Bering-Chukchi-Beaufort Seas stock	1988	7,500	6,400–9,200
Humpbacks:			
Western North Atlantic	1979–1986	5,500	2,890–8,120
Blue whales:			
Southern Hemisphere	1985–1991	460	210–1,000
Pilot whales:			
Central and eastern North Atlantic	1989	780,000	440,000–1,370,000

*Excludes Canadian east coast area. Subject to slight adjustment.
Source: International Whaling Commission (1999).

ment for individuals and up to $200,000 for organizations. Vessels may have to forfeit cargo as well.

Indigenous Alaskans are permitted to take whales and other marine mammals for subsistence purposes or for use in the manufacture and sale of native crafts. In 1998, the Clinton administration gave its approval to limited hunting of gray whales in coastal waters by the Makah Indians. The Makah reside in the state of Washington, not Alaska, and the whaling was approved for cultural, not subsistence, purposes (Northwest Indian Fisheries Commission 1998).

RECENT TECHNOLOGICAL DEVELOPMENTS AND WHALING RESTRICTIONS

Because some whale stocks have clearly rebounded and people have developed new technologies that have lowered the costs of monitoring whales and whaling, the pendulum may start to swing back toward allowing a certain amount of commercial harvesting. Commercial trading in whale products may resume as well. Some of these developments can be seen in the IWC's RMS and the current whaling plan in Norway. Recent technologies for tracking individual animals even raise the possibility of establishing complete property rights to whales.

The Revised Management Scheme

Under the IWC's RMS, it has been proposed that whaling vessels be monitored via the twenty-four satellites of the U.S. Department of Defense's global positioning system (GPS) or via ordinary radio communications. Vessels for factory-ship whaling or whaling boats that are at least twenty-two meters (seventy-two feet) long would have to be equipped for GPS monitoring. Shorter boats, such as those used by Norwegian whalers, would have to at least be able to communicate with authorities by means of shipboard radio.

The RMS is also noteworthy for its potential use of DNA technologies. Like humans and other animals that reproduce sexually, each whale has a unique genetic code that is revealed in tissue samples (Baker and Palumbi 1994; Cipriano and Palumbi 1999; Palsbøll et al. 1997). Samples can be collected by a small dart on a tether to nick the surface of the whale and retrieve a tiny piece of tissue (Lambertsen 1987), by collecting streamers of skin that peel off whales as they bump against each other during social contact (Clapham, Palsbøll, and Mattila 1993), and by taking augmentation from dead whales after either hunting or natural death. As with human fingerprints, a database of many individuals can be built over time. If products are being sold from illegally hunted whales, their DNA can be checked against the entries in the database. Norway has already developed such a database.

Earthtrust, an environmental group based in Kailua, Hawaii, has developed a same-day field test for whale species that uses a DNA polymerase primer, a chemical that highlights only the DNA characteristics that exist in a single whale stock. Using this approach, Earthtrust has been able to show the origin of meat sold in Japanese markets. Most of the sampled meat has been from (legal) Southern Hemisphere scientific hunting of minke whales, but Earthtrust has shown that some of the meat ostensibly taken from minkes has been from protected whale stocks or from dolphins (Earthtrust 1996, 1–2). The good news is that the extent to which protected whale stocks have been raided appears to have declined over time, perhaps in part because of the advent of DNA testing.

The IWC's RMS also relies on the human eye to monitor the activities of whalers. The government of each whaling nation is supposed to appoint and pay for inspectors to watch over whalers who operate in its coastal waters. There are also provisions for the possible use of international observers appointed by the IWC.

The inspectors would collect tissue samples, record data on each whale, check vessels, and report on any alleged violations of national law or IWC regulations. They would have the right to contact authorities through the ship's communications equipment. Vessels would, of course, be subject as well to surveillance by national authorities such as the U.S. Coast Guard, which has its own fleets of ships and C-130 aircraft.

NON-DNA METHODS FOR TRACKING LIVE, INDIVIDUAL WHALES

There are a few non-DNA methods for monitoring individual whales in their natural surroundings, the oldest method being a visual one. Humpback and sperm whales, for example, have a pattern of irregularities on the trailing edge of their flukes—the lobes that extend horizontally on each side of their tails. These irregularities are unique to each whale. For decades, observers have been able to verify a whale's identity by comparing pictures taken at different times.

This method is reliable, but it requires a clear photograph at each siting; the flipping of a whale's tail in the air is too transitory an event for unaided comparison of flukes. As yet, there is no information technology available to automate the task of examining a photo against a catalog of flukes. But WhaleNet, a National Science Foundation–supported project based at Wheelock College (1999) in Boston, has produced a CD-ROM with fluke photographs that have been taken since 1976 of Gulf of Maine humpback whales. Although the examination of fluke irregularities has been the most common visual way of identifying individual animals, the dorsal fin of each killer whale also has unique characteristics (SeaWorld 1999).

Another method for identifying individual whales pertains to the sounds they emit. Whale songs show characteristics by species and have the potential in at least some cases for identifying individual animals; however, the relationship between singers and the entire population of whales is not yet known. This knowledge has implications not only for the tracking of individual whales but also estimates of herd size based on detected singing.

The most promising set of technologies for tracking individual whales consists of various types of tags that can be attached to them. The word *tag* may suggest to some people a passive device like a dog tag, but whale tags are actually small data collection and broadcasting units that send out signals by radio or sound waves.

Because radio waves are by far the dominant communications technology for whale tags, the tags are always placed on the dorsal (top) surface of the whale, preferably right behind the blowhole. Salt water is opaque to radio waves at the frequencies being used, so the tag needs to be in the open air to broadcast. A dorsal tag will routinely be out of the water when a whale is at the surface. Other sides of the whale are not above water as often or as long.

Dart tags are usually fired into a whale's blubber to a depth of about ten centimeters (four inches) by using a small gun or crossbow. The instrumentation can be rigidly attached as part of the dart, or it can trail behind, tethered by a cable one to two meters (three to seven feet) long. Dermal as opposed to hypodermal tags are attached to the surface of a whale's blubber with an epoxy. A tag is mounted by means of a long pole in an operation conducted from alongside a whale in a small boat. Because such a tag does not injure a whale and because epoxy is organically inert, this type of device can stay mounted for almost three years on average.

Most whale tags broadcast in VHF signals to satellites. As noted earlier, the opacity of salt water to radio waves creates problems for transmitting signals because a dorsal tag must be able to decipher when conditions are dry enough to attempt a broadcast. Each tag is thus equipped with a saltwater switch so it transmits only while exposed to air.

There are more than 7,000 tags and instruments for various purposes in use under the Argos network of low earth-orbiting satellites and ground receiving stations. On average, a particular Argos satellite is accessible to an earth- or animal-based instrument for eight to fifteen minutes as it flies over, depending on the location of the satellite in relation to the transmitter. During that time, the satellite uses measurements of the Doppler shift of the VHF signal to calculate the position of the transmitter. This requires two or more transmissions from the satellite tag. However, position estimates based on just two messages may not be accurate. Ideally, estimates should be based on at least three messages, in which case a whale's position can be reliably established within one kilometer (0.62 mile).

Lotek Marine Technologies in St. John's, Newfoundland, manufactures ordinary (nonsatellite) radio and acoustical (sound-transmitting) tags for tracking whales. The cost per tag ranges from US$250 to US$1,100, depending on the specific components involved. Production runs are typically for 20 tags. Runs of up to 100 tags sometimes occur. Lotek reports that large cost reductions would be possible even without further improvements in technology if production could be carried out on a larger scale.[4]

It is worth noting that transponders for the satellite monitoring of taxis and other motor vehicles now cost less than $80 each. The components of a satellite tag are more sophisticated than those in transponders, but tags could still be far cheaper than they are today if there were simply a bigger market for them.

Neither Lotek nor Telonics, a leading producer of satellite tags, was able to offer estimates on tag installation costs or the costs of processing data received. Professor Bruce Mate, Oregon State University, Fisheries Department, has tagged dozens of blue whales and reports that these costs are indeed formidable. The principal problem is that a boat and crew must be sent out to tag each animal. Time, equipment, and fuel are involved. The total cost of a satellite tag, the installation, and data processing for just six months can thus amount to thousands of dollars per animal. Insofar as high-quality tags can be shot into whales with guns, one can envision times when more exotic devices will be able to tag whales from longer and longer distances. It would then not be necessary to send a boat close to the whale; however, this possibility remains speculative.

THE ECONOMICS OF WHALES TODAY

As noted, property rights are feasible only if the costs of monitoring wildlife are low in comparison to its value. Recent advances in technology have reduced the costs of monitoring whales. The value of whales has decreased in some respects while increasing in others. As suggested by the early regulations agreed to by the nations who participated in the International Convention for the Regulation of Whaling, a key product of harvesting efforts up to that point was whale oil. Petroleum- and vegetable-based oils, however, have since proven to be cheaper substitutes, and the market for whale oil has completely collapsed.[5] If it could be shown conclusively that whale oil is superior to fish oil in staving off heart disease, the demand for the former could conceivably experience a revival, but whale meat is today the biggest generator of revenues for whalers.

Norway and the Faroe Islands harvest the most whales and constitute the biggest markets for whale meat, but the harvest in each case is usually fewer than 1,000 whales annually. In Norway, whaling boats with ten or fewer people are the norm. In the Faroe Islands, people in small vessels circle behind offshore schools (pods) of pilot whales and drive them toward land with stones. The whales beach themselves and are then killed with knives. If Japan resumed full-scale commercial whaling, including the possible use of factory ships, it would probably become the largest market for whale meat, but it currently ranks third.

Although minke whales are relatively small, they would be the ones most likely to be harvested on a larger scale if whaling were to become liberalized worldwide. One place where whaling could be very lucrative is the Antarctic. If the population of minke whales in the region is more than 750,000, several thousand whales per year could be harvested on a sustainable basis with the use of modern technology. The harvesting cost per whale could be reduced significantly compared with the situation in Norway.

Fin whales represent another possible source of profits. These whales are typically more than two and one-half times as long as minkes and thus offer much more meat and blubber. They are fast swimmers, but became a popular whale to try to harvest once there existed technology capable of catching up to them. They are found all over the world, and according to the IWC, there are probably now more than 50,000 of them in the North Atlantic alone.

Greenpeace, for one, has urged that the commercial exploitation of whales center on whale-watching tours. However, in the Barents Sea, where most Norwegian whaling occurs, whale-watching tours are not particularly attractive from a commercial point of view because so few people reside in or visit the area. There are just a handful of active tour companies, and they focus on sperm whales rather than minkes. In more populated and visited locales around the world, whale watching has indeed become big business. At this writing, there are more than 375 Internet sites on Yahoo! for whale-watching excursions. Companies typically charge $35 to $100 per passenger for day trips. These tour operators might also be asked to bear some of the costs of enforcing whaling regulations.[6]

Although whale watching has become big business, perhaps the most noteworthy way in which the value of whales has increased is simply the importance that people now attach to their existence. Polling data indicate that people all over the world want whale populations to survive (Freeman and Kellert 1994; Responsive Management 1998). Irrespective of their consumption value, people are often willing to support organizations that are dedicated to conserving them, especially in wealthier countries.

But the polling data also indicate that misinformation about whales abounds. A majority of the respondents say that they believe that all whale populations are endangered and therefore oppose all commercial whaling. However, if the respondents are told that several stocks of whales are not at all threatened with extinction, most say that a limited hunt should be allowed. In other words, the data indicate that if people are assured that a species is not endangered, the value that they attach to the preservation of additional animals falls off rather sharply.

ESTABLISHING NEW RIGHTS TO WHALES

For 2001 the High North Alliance (2000) reported that Norway set a quota of 549 for the minke whale harvest in its coastal waters. The harvest involves thirty-six vessels. Individual vessels are each entitled to catch a limited number of whales annually. It should be emphasized that the entire whaling regime in Norway has not been set up to achieve efficiency in the sense of neoclassical economics, but to preserve a traditional way of life in coastal communities. Catch quotas are not tradable.

As noted, tradable quotas are appealing when it is not feasible to establish complete property rights, but some hunting can be allowed. On the other hand, freely tradable entitlements are sometimes seen as a threat to community traditions insofar as it is possible that well-capitalized outsiders would buy up scarce rights.

It would be possible to deal with such situations by vesting some whaling rights in communities rather than individuals. A community could still consider the trading of whaling rights with outsiders, but might choose not to trade if its social fabric would be thereby threatened. Tradable community development quotas (CDQs) already exist with respect to fishing rights for some Native American settlements in Alaska (De Alessi 1998, 46–47).

It would now be relatively easy in a logistical (but not a political) sense for the IWC to turn its RMP into a regime of individual transferable quotas (ITQs) and, if desired, CDQs. Each year the IWC could set catch limits for whale stocks. Whalers would have tradable shares in the total catch allowed for each stock.

The catch limit would change from time to time as new evidence emerged regarding the size of each stock. Each rights holder's share would therefore not be a fixed number of whales, but would be a percentage of the total catch allowed. For example, a whaler with a 3 percent share of the harvest of northeast Atlantic minkes would be entitled to catch 27 whales per year if the total catch limit were 900 whales annually. The best examples of already-existing plans of this sort are the fisheries of New Zealand, especially those for orange roughy (Muse and Schelle 1988).

The technology for enforcing tradable quotas for whaling would be as follows: Each vessel would be required to have a satellite hookup or a basic radio transmitter and receiver for smaller, local vessels that would enable the authorities to know the location of each whaling boat at all times. Each vessel would be equipped with a black box to record when its cannon had been fired. Vessels would be checked when they came back to port. All catches would be documented. DNA samples would be collected from each whale taken. If whale meat or blubber were being sold outside the system, authorities could verify that the DNA in question did not match any of the entries in the system's database. A small amount of whaling outside the quota system would undoubtedly occur, but it could be limited by the threat of large fines (as under current regulations).

As mentioned, the IWC's RMP seeks to keep each whale stock at between 54 and 72 percent of its historical level, with 60 percent being approximately the level corresponding to the maximum sustainable yield (MSY) of whales. The total catch limit for each stock would depend on an estimate of the stock's MSY.

The Possible Role of Bioeconomic Models

Although a quota system could conserve whale stocks, it leaves open the question as to what the right size is for each stock. The stock level that would generate the maximum sustainable yield of whales is a biological, as opposed to an economic, concept. From an economic perspective, the MSY stock level is not necessarily optimal, because the value of a whale may change over time. A particular whale may be more valuable today than the same whale will be one year from now (or vice versa). If so, it is economically preferable to harvest the whale today even if such activity would cause a stock to fall below its MSY level.

Researchers have devised bioeconomic models (Clark 1973, 1976a, 1976b) to analyze in a dynamic context the characteristics of an optimal stock. Based on a bioeconomic model, one possible management strategy would be to estimate the intrinsic growth rate and other parameters of the model and determine the optimal size for each stock. Amundsen, Bjørndal, and Conrad (1995) thus estimated that it would be optimal for the North Atlantic stock of minke whales to have 52,000 to 82,000 adult animals. Harvesting quotas could then be established that are consistent with this amount. The desired stock, and in turn, the harvesting quotas could be adjusted as new parameter estimates were obtained. As the above estimate suggests, however, this approach still leaves the precise size of the optimal stock uncertain.

Tradable Shares in the Total Stock

A possible way to cope with this problem is to define tradable harvesting rights to the stock of whales itself and then let market forces discover the right number of animals (Alchian 1950; Hayek 1978). One could begin with a conservative (i.e., lower bound) estimate of the size of the stock. Rights to whales could be made proportional to historical catch data (insofar as they exist) or according to other criteria.[7] The sum of the stock rights would correspond to the estimated size of the stock. As noted in the case of tradable harvesting quotas, people would have shares in the total stock, but not rights to a fixed number of whales. If people did not exercise harvesting rights in any particular year, the stock would grow, people would be entitled to take additional whales in the future, and the whale-value of their stock rights would increase.

As a simple illustration, suppose that the size of the stock of eastern Pacific gray whales were estimated to be 30,000 animals. If someone began with a 2 percent share of this stock, he would have harvesting rights to 600 whales. If he proceeded to kill 600 animals, he would no longer have any harvesting rights. However, if the intrinsic annual growth rate of the stock were estimated to be 3 percent, and if he did not harvest any whales during the year, he would then have harvesting rights to 618 (600 multiplied by 1.03) whales. Adjustments to the whale-value of the person's stock rights would be necessary following periodic reevaluations of the total size of the stock. Implicit in any trading of stock rights would be market assessments of how the estimated size of the stock would change in the foreseeable future.

A basic, and probably decisive, argument against such a regime is the great uncertainty about the sizes of most whale stocks. In some cases, the outlines of a discrete stock are not even well defined. Stock estimates are also subject to large and, some would say, arbitrary revisions, and many years are required to undertake them. For those especially concerned about the longer-term viability of whale populations, an annual ceiling on the total catch allowed for each stock therefore would probably inspire more confidence, at least for the time being.

Ownership of Individual Animals

Further improvements in technology could eventually reduce tagging costs and make viable the commercial tracking of individual members of various whale stocks. In this eventuality, people could have tradable property rights to individual whales. People could then increase their individual stocks by tagging any calves traveling with their mature female whales.[8] The whale-value of people's stock rights would not depend on periodic estimates of the total stock size and its rate of growth.

A conceivable way to economize on tagging costs is to mark only a few members of a pod of whales. It is likely that unmarked members would still be nearby because of the nature of a pod. Some of the toothed species such as sperm, killer, and pilot whales do indeed travel in groups. Although this strategy would save on the tagging and tracking costs of the pod as a whole, it would leave untagged animals more vulnerable to poaching. Moreover, the baleen (toothless) whales seldom travel in groups.[9]

ADDITIONAL COASEAN CONSIDERATIONS

We have seen that a workable plan of tradable stock rights is probably still some time away. However, it is instructive to consider further the form that such a course of action might take as the relevant technology continues to improve. According to a simple version of the Coase (1960) theorem, if the costs of transacting were very low, it would not much matter for the allocation of resources how stock rights were initially assigned. Trading ensures that rights would be put to their highest-valued uses, whatever they might be. If particular whales have more value as a source of pizza toppings than as the subject of a tourist's photo session, whale-watching companies would be encouraged to sell any rights that they might have to whalers. If, on the other hand, particular whales have great value simply as magnificent creatures whose existence is to be nurtured and cherished, conservation groups would tend to end up with the rights to those whales.

Reality is not always simple, however. Transaction costs are sometimes high. In particular, there is a free-rider problem with respect to the collection of donations to conservation groups. People may value their activities, but they may often refrain from making donations in the hope that others' donations will suffice to accomplish the group's goals. The amount received in donations may then be suboptimal. The conservation groups could thus have difficulty in achieving their goals even if people really do value their activities more than commercial uses of the resources in question.

A major concern for any regime of tradable stock rights is to ensure that stock extinction does not occur. Conservation groups could conceivably buy up the stock rights of any whalers who were indeed intent on harvesting an endangered group of whales, but they would have a difficult time doing so if they could not collect large amounts of donations.

Among minke whales in the North Atlantic, there was no tendency for extinction to occur even in the years when the stock existed as an open-access commons (Amundsen, Bjørndal, and Conrad 1995). The stock of whales was driven below its optimal level, but the costs of locating and catching the remaining whales exceeded the revenues that could be thereby obtained. On the other hand, the case of the blue whale during the heyday of factory whaling in the Antarctic—in which the number of blue whales there was probably driven below a thousand—argues for more caution. To be sure, the blue whales also existed as an open-access commons, but if their population growth rate is low, bioeconomic models warn that a carelessly devised system of tradable stock rights would not guarantee their survival.

A way to deal with the possibility of extinction under a plan of tradable stock rights is to give significant numbers of rights at the outset to conservation trust organizations. An example of such an organization is the Center for Coastal Studies (CCS), a private nonprofit group in Provincetown, Massachusetts (www.ccs.org). This organization uses satellite and ordinary radio tags to monitor the status of humpback and North Atlantic right whales from the coast of Maine to the tip of Florida. If a whale becomes trapped in fish netting or buoy lines or is otherwise in trouble, CCS is authorized by the National Marine Fisheries Service to engage in a rescue mission. Such a mission typically involves disentangling the whale from the material in which it was trapped.

In keeping with the spirit of the Coase theorem, a CCS endowed with stock rights might occasionally have an incentive to sell the rights to one or two animals if their value for some other purpose were sufficiently high (and CCS could therefore get a very good price for them). However, it could legally sell rights to whales only if such a transaction would arguably result in a net enhancement of its ability to carry out its mission to conserve whale populations. For example, the proceeds of a sale might help finance the purchase of new monitoring or rescue equipment, so that giving up one or two whales from an abundant stock now could help save several others from an endangered stock at a later time.

Public trust organizations might be established to supplement private groups. Entire stocks of whales could be turned over to conservation trusts, making it possible to have some trading of rights to whales without the tagging of all the individual animals. It should be pointed out, however, that a conservation trust does not involve complete property rights if there are not tradable ownership rights to the organization itself. There is thus not a complete market for corporate control, and a certain amount of inefficiency may be endemic to the trust's operation. A for-profit corporation that is poorly managed can, by contrast, be bought out by an entrepreneur who offers the prospect of improving its performance.

SUMMARY AND CONCLUSION

Changes in technology have interacted with the institutional environment for whaling in dramatic fashion over time. In the past, advances in technology resulted in the depletion of whale populations at a time when they existed as an open-access commons. It thus became plausible to regulate the whaling industry.

Unfortunately, the International Whaling Commission proved to be incompetent. First, it countenanced overwhaling. Now it mandates underwhaling. Some stocks of whales are now clearly abundant, but the IWC's moratorium on commercial whaling continues. Recent advances in technology have helped to make possible a relaxation of restrictions on whaling, as seen in Norway's management of sustainable whaling and in the proposed RMS of the IWC. But neither Norway's regime nor the IWC's provisions for a possible return to larger-scale commercial whaling involve tradable whaling rights.

An enforceable system of tradable rights is workable with the use of current technology. From a logistical perspective, a system of transferable quotas (ITQs or CDQs) could easily be made a part of an IWC-sponsored whaling plan. The available technologies for enforcing the rules of such a plan include GPS satellites and DNA testing.

From an economic perspective, existing IWC provisions under the RMP are excessively conservative insofar as they would attempt to keep whale stocks at a level that corresponds to each stock's maximum sustainable yield of whales. MSY is a biological concept, not an economic one. As an alternative to an MSY regime, the economically optimal size of each whale stock could be estimated through the use of a bioeconomic model, and tradable quotas could be established that are consistent with these optima.

A more radical approach is to establish tradable stock rights and allow market forces to discover the appropriate level for each whale stock. Regulatory authorities would not control the total harvest of whales, but they would still have a role to play in adjusting the whale-value of people's stock rights according to periodic estimates of the size of each whale stock and its intrinsic rate of growth. Although tradable stock rights could conceivably be established, estimating the size of each stock and its growth rate is problematic, so much so that the plan would be widely regarded as arbitrary. The uncertainty involved would also skew the decisions of profit-seeking whalers toward near-term exploitation of their harvesting rights.

It is still not feasible with current technology to have widespread ownership of live, individual whales on a commercial basis. Satellite tagging is still far too expensive in comparison to the commercial value of most whales, especially smaller species like the minke. There would be more commercial interest in various types of tagging if the market for whale oil had not collapsed.

Whole stocks of whales could be turned over to conservation trusts without the tagging of all the individual animals. These organizations could sell whaling rights as long as the revenues from such sales arguably enhanced their conservation efforts. Conservation trusts, however, do not involve complete property rights.

If a significant commercial demand for harvesting them remains, further advances in technology could indeed make it feasible one day for there to be widespread ownership of individual whales. Tags or other such devices will cost in real terms only a fraction of what they cost today. Installation costs will also decline if people no longer need to be right next to a whale in order to tag it. With widespread tagging of individual animals, the whale-value of people's stock rights would not depend on periodic estimates of the size of each whale stock or its intrinsic rate of growth, just as it is not necessary to estimate the total population or growth rate of each cattle species in order to articulate people's property rights to cattle.

If secure private property rights to individual whales could be established, the role of regulatory authorities could recede. Markets would determine outcomes. Whales would be like swimming cattle. Our story would be complete, except for an account of all the politics that would be necessary along the way.

NOTES

1. Quoted in Watson (1993).
2. This point was not made by Lueck (1991) or Anderson (1998), but it is consistent with their theoretical framework.
3. Ray Gambell, secretary, International Whaling Commission, e-mail communication, July 14, 1999.
4. We are particularly indebted to Padraic O'Flaherty, Marine and Freshwater Application sales representative, Lotek, for providing us with information about the company in an e-mail dated September 3, 1999.
5. The sperm whale is most noteworthy here because of the large quantity of high-quality oil that can be had from its massive head. The sperm whale is also a source of ambergris, commonly referred to as Neptune's treasure. This waxy substance appears to be a by-product of the animal's digestive system and is prized by the perfume industry (Clapham 1997, 98). But it is difficult to assess the market for this substance in light of its illicit nature. Synthetic substitutes for it now exist.
6. Note that if a whale-watching organization were given the rights to a group of whales, there would be a possible free-rider problem if others might still be able to venture out and see them. The rights of the organization would thus, for example,

have to include a stipulation that rival tour boat operators keep a certain distance from the whales.

7. Note that if people are aware in advance that rights are likely to be allocated on the basis of historical catch data, they have an incentive to hunt animals now in order to bolster their claims.

8. It should be noted that captive breeding and rearing of whales is rarely desirable at present because of the quantity of food that they require. According to Six Flags Marine World (Vallejo, California), which stages whale and dolphin shows, a killer whale consumes 100 to 150 pounds of fish per day. The monthly food bill for one whale is $2,000 to $3,000. Other expenses include veterinary care and the amortized costs of providing a tank and water filtration system for the animals.

9. In the case of pilot whales off the Faroe Islands, property rights could be assigned on a territorial basis. At present, whales can be harvested in any of twenty-two bays under the supervision of whaling sheriffs (High North Alliance 1995). It would be possible to assign exclusive property rights to the whales traveling into particular bays. However, it should be emphasized that whaling in the Faroe Islands is done on a subsistence basis. In such circumstances, communal management may be desirable as a way of pooling the risks to people's food supply that are involved in the likelihood that the harvest in some years will be rather low.

REFERENCES

Alchian, Armen A. 1950. Uncertainty, Evolution, and Economic Theory. *Journal of Political Economy* 58(June): 211–21.

Amundsen, Erik S., Trond Bjørndal, and Jon M. Conrad. 1995. Open Access Harvesting of the Northeast Atlantic Minke Whale. *Environmental and Resource Economics* 6(September): 167–85.

Anderson, Terry L. 1998. Viewing Wildlife through Coase-Colored Glasses. In *Who Owns the Environment?* ed. Peter J. Hill and Roger E. Meiners. Lanham, Md.: Rowman & Littlefield Publishers, 259–82.

Anderson, Terry L., and Donald R. Leal. 1997. *Enviro-Capitalists: Doing Good While Doing Well.* Lanham, Md.: Rowman & Littlefield Publishers, 103.

Aron, William, William Burke, and Milton Freeman. 1999. Flouting the Convention. *Atlantic Monthly*, May, 22–29.

Baker, C. S., and S. R. Palumbi. 1994. Which Whales Are Hunted? A Molecular Genetic Approach to Monitoring Whaling. *Science*, 9 September, 1538–39.

Cipriano, Frank, and Stephen R. Palumbi. 1999. Genetic Tracking of a Protected Whale. *Nature*, 28 January, 307–08.

Clapham, Phillip. 1997. *Whales of the World.* Stillwater, Minn.: Voyageur Press.

Clapham, Phillip J., Per J. Palsbøll, and David K. Mattila. 1993. High-Energy Behaviors in Humpback Whales as a Source of Sloughed Skin for Molecular Analysis. *Marine Mammal Science* 9(April): 213–20.

Clark, Colin W. 1973. Profit Maximization and the Extinction of Animal Species. *Journal of Political Economy* 81(August): 950–60.

———. 1976a. A Delayed-Recruitment Model of Population Dynamics, with an Application to Baleen Whale Populations. *Journal of Mathematical Biology* 3: 381–91.

———. 1976b. *Mathematical Bioeconomics: The Optimal Management of Renewable Resources.* New York: John Wiley & Sons.

Coase, Ronald H. 1960. The Problem of Social Cost. *Journal of Law & Economics* 3(1): 1–44.

De Alessi, Michael. 1998. *Fishing for Solutions.* London: Institute of Economic Affairs.

Earthtrust. 1996. ET's DNA Work Continues to Set the World Standard. Available: www.earthtrust.org/prezltr2.html. Cited October 23, 2000.

Freeman, Milton R., and Stephen R. Kellert. 1994. International Attitudes to Whales, Whaling, and the Use of Whale Products: A Six-Country Survey. In *Elephants and Whales: Resources for Whom?* ed. Milton R. Freeman and Urs P. Kreuter. Basel, Switzerland: Gordon and Breach Science Publishers.

Gambell, Ray. 1993. International Management of Whales and Whaling: A Historical Review of the Regulation of Commercial and Aboriginal Subsistence Whaling. *Arctic* 46(June): 97–107.

Hayek, F. A. 1978. Competition as a Discovery Procedure. In *New Studies in Philosophy, Politics, Economics, and the History of Ideas.* Chicago: University of Chicago Press, 179–90.

High North Alliance. 1995. *Pilot Whaling in the Faroe Islands.* Available www.highnorth.no/Library/Hunts/Faroe_Islands/pi-wh-in.htm. Cited October 23, 2000.

———. 2000. Norway: Further Decrease in Whaling Quota. December 1. Available www.highnorth.no/. Cited January 29, 2001.

International Whaling Commission. 1999. Revised Management Scheme: Draft of the Revised Chapter V of the Schedule Proposed by the Government of Japan. Papers of the 51st Annual Meeting of the International Whaling Commission, Grenada, May 24–28.

Jackson, Gordon. 1978. *The British Whaling Trade.* London: Adam and Charles Black.

Lambertsen, Richard H. 1987. A Biopsy System for Large Whales and Its Use in Cytogenetics. *Journal of Mammology* 68(May): 443–45.

Lueck, Dean. 1991. Ownership and the Regulation of Wildlife. *Economic Inquiry* 29: 249–60.

Motluk, Alison. 1996. Blood on the Water. *New Scientist* 150(June 22): 12–13.

Muse, Ben, and Kurt Schelle. 1988. *New Zealand's ITQ Program.* Alaska Commercial Fisheries Entry Commission Paper 88-3. Juneau, June.

New Encyclopaedia Britannica. 1991. Whaling. 15th ed. Chicago: Encyclopaedia Britannica.

Northwest Indian Fisheries Commission. 1998. Makah Management Plan for Makah Treaty Gray Whale Hunting for the Years 1998–2002. Available www.nwifc.wa.gov/whaling/whaleplan.html. Cited May 15, 2000.

Palsbøll, Per J., Judith Allen, Matine Bernbé et al. 1997. Genetic Tagging of Humpback Whales. *Nature*, August 21, 767–69.

Responsive Management. 1998. Knowledge of Whales and Whaling and Opinions of Minke Whale Harvest among Residents of Australia, France, the United Kingdom and the United States. Research report. Harrisonburg, Va., April.

SeaWorld. 1999. Killer Whales: Habitat and Distribution. Available http://jj.seaworld.org/killer_whale/habdiskw.html. Cited May 30, 2000.
Spencer, Leslie, and Jan Bollwerk (with Richard Morais). 1991. The Not So Peaceful World of Greenpeace. *Forbes*, November 11, 174–77.
Starbuck, Alexander. [1878] 1964. *A History of the American Whale Fishery from Its Earliest Inception to the Year 1876.* New York: Argosy Antiquarian.
Tonnessen, J. N., and A. O. Johnsen. 1982. *The History of Modern Whaling.* London: C. Hurst and Co.
U.S. Department of State. 1946. *The International Convention for the Regulation of Whaling, 1946.* December 2. Available http://ourworld.compuserve.com/homepages/iwcoffice/convention.htm. Cited October 23, 2000.
Watson, Paul. 1993. An Open Letter to Norwegians. January 8. Available www.highnorth.no/Library/Movements/Sea_Shepherd/le-to-th.htm. Cited October 23, 2000.
Wheelock College. 1999. *WhaleNet Guide to Whales and Marine Mammals of the North Atlantic Ocean* (CD-ROM). Boston: Wheelock College.

7

Feasibility of Contaminant Source Identification for Property Rights Enforcement

Anna M. Michalak

Environmental contamination issues span a variety of natural and anthropogenic chemicals and the full range of possible media, ranging from oceans to the atmosphere. Despite significant advances in the scientific understanding of contaminant transport, however, polluters' liability for damage caused to downstream victims has been difficult to determine and enforce. This paper explores the feasibility of using scientific methods of contaminant source identification as a mechanism for enforcing the property rights of individuals affected by contamination.

PROPERTY RIGHTS AND CONTAMINATION

The protection of rights in property lies at the heart of the American constitutional system (Marzulla and Marzulla 1997, 1–9). For the discussion presented here, property rights will be defined as the rules that specify who has access to resources and the authority to derive value from them. These rules may be formal, such as state enforced laws, or informal, such as customary rights. In cases of environmental contamination, the question is who has the rights to the disposal medium, be it air, water, or soil. It is a matter of competing uses. Does the party generating the waste have a right to dispose into the medium, or do others have a right to enjoy the benefits of an unspoiled medium? The question of such conflicting property rights was addressed by Coase (1960), who concluded that well-defined and enforced property rights are a prerequisite to market solutions to conflicting uses, assuming that transaction costs are not prohibitive. Therefore, the clear definition of property rights and their enforcement in legal settings can help bring about a bargaining solution to contamination problems, and help avoid the tragedy of the commons.

Two clear ways to minimize transaction costs in environmental contamination cases are to conclusively identify the source of the contamination and to have a legal system that enforces property rights by assigning liability for damage caused to other property owners. When these two criteria are not met, a system of property rights for managing environmental contamination cases cannot work. This chapter focuses on the feasibility of attaining the first criterion, given the current level of scientific knowledge and environmental policy system.

When the property rights of polluters and those being harmed by the contamination are enforced, property owners affected by contaminant migration from an adjoining property can defend their rights based on the legal theories of nuisance, trespass, and, in some cases, strict liability. Under the theory of strict liability, the responsible party is held accountable without regard to fault. Modern courts have applied strict liability to activities "characterized as 'perilous,' 'ultra or extra-hazardous,' or 'abnormally dangerous'" (*Woodrow Sterling et al. v. Velsicol Chemical Corporation* 1986). The theories of trespass and nuisance are more broadly applicable. A trespass is defined as an invasion of the interest in the exclusive possession of land as by entry upon it, whereas a nuisance is an interference with the interest in the private use and enjoyment of the land, and does not require interference with the possession (*Lopardo v. Fleming Cos.* 1996). In order for these legal theories to be enforceable, however, the source and the effect of the contamination must be determined. Furthermore, the policy setting must enforce property rights by imposing liability for the effects of contamination.

To understand how these ideas fit into recent contamination cases, a historical review of the treatment of contamination issues is required. In considering industrial pollution, the courts have faced a conflict between fundamentally opposed principles of property rights. Did industries have a legitimate right to discharge pollutants as an unfortunate but unavoidable by-product of the material benefits they provided, or did those who suffered the consequences have a right to pure air and water undefiled by industry? Traditional British common law supported the latter position, allowing landowners to block development on grounds that it interfered with their "quiet enjoyment" of the "natural use" of their estates. The common law in America, in contrast, gradually changed to favor the rights of those who first put natural resources to productive use (Andrews 1999, 126–29; Horwitz 1988, 51–52). In effect, this evolution replaced the principles of common law rights with doctrines of economic balancing. These precedents are similar in concept to more recent legal arguments for balancing of the economic interests of polluters against those of parties being affected by the contamination, rather than upholding clearly specified property rights (Andrews 1999, 126–29).

Since the 1970s, the results of environmental policy have been mixed. In the area of water pollution policy, for example, with some important exceptions, progress has been made on reducing point-source discharges of conventional pollutants, but the cost has been enormous. U.S. businesses, government, and individuals have spent more than $300 billion (in 1984 dollars) since 1972 for water pollution control (Anderson and Leal 1991, 142). Marzulla and Marzulla estimated that, overall, environmental regulation annually costs Americans between $150 and $300 billion. The federal government, however, annually recovers only about $1 billion in fines, cleanup costs, and injunctive relief in suits against American companies (Marzulla and Marzulla 1997, 93). These statistics raise questions about the economic logic of current methods of addressing contamination.

Contaminant Source Identification

A Coasean bargaining solution with well-specified property rights would be more feasible if the source(s) of a particular contaminant could be determined at low cost. First, the receptor's burden of proof would be reduced because only the contaminant's harmful effect and not its origin would remain to be demonstrated. Individuals being affected by contamination could more easily enforce their property rights against trespass and nuisance caused by contaminants. Second, individuals engaging in activities that result in some contamination could use source identification to show that their activities are not the source (or not a significant source) of observed harm and could continue their economic enterprise. Third, if receptors affected by a given contaminant were informed of the identity of the responsible party, an optimal level of emissions or remediation could be negotiated, with appropriate compensation for the affected parties. Especially in cases in which the polluter is easily identifiable or the damage is concentrated on a single individual or a small group, there is an incentive to arrive at an optimal level of pollution. Last, if the individual contributions of several sources could be determined, the free-rider problem would be eliminated because the contribution of each polluter would be known and the environmental initiatives of one industry would not benefit neighboring industries. In essence, the availability of effective methods of contaminant source identification, followed by their recognition in legal and political settings, would allow parties affected by contamination to enforce their property rights to clean water, air, and soil. This, in turn, would force responsible parties to recognize the property rights associated with their emissions and to be accountable for their costs.

The idea of using contaminant source identification as a method of enforcing property rights has been proposed. Stroup (1990, 208) suggested that chemicals that might escape into the water or air could be branded by radioisotopes, dyes, or other means to help identify their source if later they escape, appear where they are unwanted, and cause damage. Former Environmental Protection Agency (EPA) analyst Fred Smith explored the possibilities of available technologies:

> Tracers (odorants, coloring agents, isotopes) might be added to pollutants to ensure the damages were detected early where the costs of reduction were lower. . . . There are exotic technologies that might well play a fencing role even for resources as complex as airsheds. . . . If that system were combined with a system under which each nation adopted some fingerprinting system to identify its major greenhouse gases . . . , it would become possible to trace pollution to its source and thus make the polluters pay. Note that most developed nations do participate in "labeling" high explosives manufactured in their countries as part of a worldwide anti-terrorist program. (quoted in Anderson and Leal 1991, 166)

If contaminant source identification is to be used as a property rights enforcement tool, the underlying scientific methods must be reliable enough to be convincing. This chapter examines current types of methods available for contaminant source identification and case studies illustrating the advantages and limitations of the utilized technology. The methods are then compared based on their scientific merits and their applicability in the current policy setting. Finally, changes in policy focus that may help ease the application of source identification technologies are discussed.

AVAILABLE METHODS FOR SOURCE IDENTIFICATION

Many methods are available for identifying contaminant sources. These methods differ in their ranges of applicability in terms of contaminants and media, the level of confidence associated with their results, and a variety of other factors. Despite such wide variability, contaminant source identification methods can be subdivided into three categories: compositional analysis, the use of naturally occurring or introduced tracers, and methods that derive conclusions made from the contaminant distribution itself.

Compositional Analysis

Compositional analysis methods identify specific components of a given contaminant. The source of the contamination can then be determined if the presence or abundance of these components correlates predictably with contaminant sources. These methods can generally be subdivided into methods that analyze the molecular composition of compounds and those that determine their isotopic composition.

The first form of compositional analysis, based on the molecular composition of a given contaminant, is useful in cases in which the contaminant consists of a source-dependent mixture of various substances. This can occur either naturally or through a manufacturing process. For example, compositional analysis is commonly used for the identification of sources of oil contamination because oil is a mixture of various molecular components and because the fraction of each component varies naturally with the source of the crude oil. Specific methods include infrared spectroscopy analyses, thin-layer chromatography, fluorescence, gas chromatography, and gas chromatography/mass spectrometry. Several methods can be used in conjunction to give statistically independent parameters. When these independent methods agree, the statistical confidence is overwhelming (Bentz 1978; Mansuy, Philp, and Allen 1997). In the case of atmospheric particulates, the particulates differ in their elemental composition due to the different industrial processes that create them, and similar techniques can be used for source identification. For example, laser microprobe mass analysis can be used to detect the inorganic and organic molecular constituents of the particulates. The sensitivity of the mass analysis is such that a number of impurities can be detected with very low detection limits (Surkyn, de Waele, and Adams 1983).

In many cases, sources cannot be distinguished by the molecular composition of the contaminant, either because this composition has changed as a result of processes such as evaporation and biodegradation or because the various potential sources are emitting contaminants that are not sufficiently distinct to allow for differentiation. For example, the effects of weathering can strongly modify the fingerprints and parameters used to correlate an oil with its source. Even in cases of extreme weathering, however, the isotopic compositions of the individual compounds are not extensively affected, and isotopic analysis using gas chromatography/isotope ratio mass spectrometry can be used as a complementary correlation technique (Mansuy, Philp, and Allen 1997).

With simple atmospheric contaminants such as sulfur dioxide or carbon monoxide, isotopic analysis can be effective in source identification because, although all sources emit contaminants with identical molecular compositions, the contaminants sometimes exhibit distinct isotopic signatures. But the number of sources that can be examined is limited due to the small number of isotopically distinct contaminant signatures (Bergamaschi et al. 1998).

The possibilities and limitations of compositional analysis are illustrated in three case studies in which these methods were applied both for legal and scientific purposes.

Lead as a Gasoline Additive

Lead is unique in that it has variations among mining districts in the relative abundance of its stable isotopes (Rabinowitz 1995; Rabinowitz and

Wetherill 1972). Because the relative amounts of its four isotopes can change only when the lead is mixed with a different lead, the isotope ratios provide a natural tracer for following a particular batch of lead. The covariance of the ratios usually allows for only two sources to be considered, but potential sources can often be ruled out (Rabinowitz 1995).

Isotopic analysis was an important part of the evidence presented in *Ethyl Corporation et al. v. Environmental Protection Agency* (1976), which involved various manufacturers of lead additives and refiners of gasoline appealing the promulgation of low-lead regulations by the EPA. At the time, approximately 90 percent of gasoline manufactured in the United States contained lead additives, and lead particulate emissions from gasoline engines accounted for about 90 percent of the lead in the air. Despite apparent reasons for concern, hard proof of any danger caused by lead in automotive emissions was difficult to obtain, partly because of the multiple sources of human exposure to lead; the extent to which inhaled lead actually enters the body was vigorously contested by petitioners and lay at the heart of the appeal. The court found in favor of the EPA based on presented experimental calculations, epidemiological studies, and clinical studies demonstrating the significance of lead absorption from ambient air by the human body.

The clinical lead isotope study presented evidence of the relative importance of lead absorption by inhalation versus ingestion. Since lead isotope ratios in food and air were identical, the dietary lead ratio of the study's subjects was altered by giving the volunteers a low-lead diet supplemented with a lead-204 tracer. Thus, the ingested and inhaled lead could be distinguished and traced (Rabinowitz, Wetherill, and Kopple 1977). The study concluded that approximately 30 percent of blood lead was a product of airborne lead.

Source Identification for Atmospheric Methane and Carbon Monoxide

Methane (CH_4) and carbon monoxide (CO) are considered two of the most important trace gases in the atmosphere, mainly due to their large influence on atmospheric chemistry, and, in the case of methane, its significant impact on the earth's radiative budget (i.e., the greenhouse effect). Considerable methane emissions originate from Eastern European countries and the area of the former Soviet Union. In particular, West Siberia's gas fields, containing about 35 percent of world natural gas reserves and an extensive infrastructure for exploitation and transmission, represent a potentially significant source, which had not been quantified until recently. In addition, the vast Russian subcontinent harbors other globally important sources of CH_4 and CO, including areas of tundra and wetland biological activity, forest and steppe burning, and industrial emissions.

The isotopic composition of methane is significantly different for the three principal methane production processes: biogenic (bacterial) methane formation, thermogenic formation, and incomplete combustion of biomass or fossil fuels. In the case of carbon monoxide, technological sources (transport, combustion, or industrial processes) exhibit different signatures from CO produced from biomass burning, which are themselves distinct from CO produced via CH_4 oxidation far away from CH_4 sources.

An international team of scientists undertook a study aimed at identifying the sources of Siberian CH_4 and CO by analyzing their isotopic compositions. Continuous CH_4 and CO measurements were taken along the 5,821-mile Trans-Siberian railroad, leading from Moscow to the port city of Vladivostok at the Sea of Japan. The investigators were able to unambiguously show that biomass burning caused the increase in CO concentrations observed in some areas. Furthermore, the researchers found West Siberia ambient methane to be typical of biogenic sources and significantly different from thermogenic methane, suggesting that a dominant fraction of the methane originates from wetlands and not from natural gas mining and transport operations (Bergamaschi et al. 1998). As in the previous case study, the scientists were able to reach conclusions because few potential sources were being considered and these sources resulted in isotopically distinct versions of the examined gases.

PCBs in Sediments

Although polychlorinated biphenyls (PCBs) have been manufactured for industrial uses since 1929, they were not identified as an environmental threat until 1966. Commercial PCBs are complex mixtures of varying congener concentrations or weight percents. Monsanto Chemical Company, the sole PCB producer in North America from 1929 to 1975, had sold these mixtures under the trade name Aroclor. PCBs are well known for their chemical stability, low volatility, nonflammability, and high dielectric constant. Although these properties made PCBs a preferred choice for many industrial applications, they also account for widespread PCB distribution and accumulation in ecological systems.

Principal component analysis, a statistical method, was used to identify and apportion PCB sources in the Milwaukee Harbor Estuary sediments, in Wisconsin. The method is based on an analysis of sediment cores from the estuary bottom. The concentrations of eight distinct PCB congeners were measured in the cores. By comparing the congener concentrations in the samples with the corresponding concentrations known to be present in various Aroclors, the source contributions of the individual Aroclors, in principle, could be determined. Use histories of the Aroclors then could be taken into account in order to identify physical sources of the sediment contamination.

For the Milwaukee Harbor Estuary study, contributing Aroclors were identified for the sampled cores. Profiles of Aroclor 1242 and 1016, however, dominated the same low-molecular-weight region and shared most of the congener pattern characteristics, thereby complicating the source separation. This incomplete source resolution led to the calculated presence of Aroclor 1016 in the bottom (oldest) layers, although Aroclor 1016 was not produced until 1970 and therefore could not have contributed to the PCB contamination of these older sediments (Rachdawong, Christensen, and Chi 1998). Compositional analysis was not completely effective due to the large number of potential sources and the similarity of their contaminant compositions.

Tracer Use

Another category of source identification technologies is the use of tracers. The composition of a given contaminant can be used to determine its source only when the potential sources exhibit chemically distinct versions of the contaminant. Even in cases in which such analysis is not possible, however, tracer compounds discharged with the contaminant can be used to identify contamination sources. The tracer compound can be a chemical that is naturally present in the discharge or a compound added to the source for the specific purpose of tracing a contaminant. In either case, care must be taken to select a tracer with transport characteristics similar to those of the contaminant being traced in order to avoid misidentifying sources.

Analyses based on the use of naturally occurring tracers are particularly useful when a long delay is observed between the time of contaminant release and its detection at the location of interest, such as when it travels very slowly or must be tracked over long distances. In such cases, an introduced tracer would take an unacceptable amount of time to reach receptors from the start of the tracer release. In groundwater contamination cases, naturally occurring tracers are often used because transport is typically very slow, and the time scales involved in the transport can be years or decades. For example, groundwater discharge to the ocean can be traced using radon (^{222}Rn) and methane (CH_4). This is possible because both ^{222}Rn and CH_4 are naturally occurring trace gases present in groundwater at concentrations that are elevated by several orders of magnitude relative to seawater (Cable et al. 1996). A second example involves phage typing of *Escherichia coli*, one of the most ubiquitous bacteria in existence, as a fingerprinting tool to identify the source(s) of fecal coliform contamination of drinking water wells (Rusin et al. 1992). Subsurface environments, however, offer extensive possibilities for differential degradation, sorption, and dispersion. As a result, the identification of tracers that behave similarly to a given contaminant is difficult in most cases of groundwater contamination.

When contaminants are tracked in the atmosphere or surface waters, the affected area can be enormous, which also increases transport times. Such is the case when tracers are used to track atmospheric particulate matter. Several volatile organic compounds have been identified as candidates for naturally occurring tracers of fine-particulate carbon and extractable organic matter from mobile atmospheric sources because they exhibit similar transport characteristics to the bulk of source emissions (Zweidinger, Stevens, and Lewis 1990). In surface water, one application of naturally occurring tracers is the use of uric acid—it is present at significant levels in untreated sewage and not normally detected in unpolluted waters; therefore, it can be used to track sewage effluents discharged into fresh and saline waters (Brown et al. 1982). Another application of tracer use is the measurement of fluorescence and salinity. These tracers are useful for determining the behavior of constituents entering estuarine systems and tracking their dispersal (Kucklick and Bidleman 1994). For example, natural fluorescence can be used to track the dispersal of organic contaminant plumes such as herbicides (Rachdawong, Christensen, and Chi 1998). In this case, an additional reason for relying on naturally occurring tracers is that the contamination generally results from nonpoint sources, which would complicate the use of introduced tracers because the exact distribution and location of the contaminant flux are unknown.

When there is no suitable naturally occurring tracer, a unique substance can be added under controlled conditions to a source to achieve a more stable and predictable concentration at receptor sites relative to naturally occurring tracers (Ali et al. 1997). Tracer introduction is not possible for some applications, however, because of the transport time scales involved. Due to processes such as sorption and deposition that can retain a tracer within the environment for long periods, a tracer cannot always be reused because residual concentrations would limit the significance of subsequent results.

Naturally occurring and introduced tracers have been successfully used by environmental agencies; three examples of these applications are presented in the following case studies.

Visibility Impairments at the Grand Canyon National Park

The case of *Central Arizona Water Conservation District et al. v. United States Environmental Protection Agency* (1993) involved petitioners contesting regulations promulgated by the EPA to partially remedy visibility impairments at the Grand Canyon National Park (GCNP). The regulations required a 90 percent reduction in sulfur dioxide (SO_2) emissions at the Navajo Generating Station (NGS), a power plant situated approximately twelve miles from the Grand Canyon.

The regulations were promulgated as part of a national effort to control visibility impairments "that can be traced to a single existing stationary facility or small group of existing stationary facilities" (*Central Arizona Water Conservation District et al. v. United States Environmental Protection Agency* 1993, 3).

The EPA refers to this as a "reasonably attributable" impairment, defined as "attributable by visual observation or any other technique the State deems appropriate" (3–4). The states, or EPA, thus have broad discretion in determining how and whether impairment may be attributed to an individual source.

The regulations were based on results of the Winter Haze Intensive Tracer Experiment (WHITEX), a winter visibility attribution study, conducted by the National Park Service (Pitchford et al. 1999). WHITEX involved the release of a unique tracer gas, deuterated methane (CD_4), from the NGS (Anderson and Leal 1991; Crawford 1990). Based on the analysis of the experimental results, the National Park Service estimated that the NGS was responsible for 70 to 80 percent of the sulfates measured at Hopi Point (which is in GCNP) during WHITEX.

Salt River Project, the operators of the NGS, conducted a second tracer study, the NGS Visibility Study, to provide input to the rule-making process of the EPA regarding the generating station's SO_2 controls. Perfluorocarbon tracers were released from each of the three stacks of the NGS. This study also indicated a contribution of NGS emissions to haze in GCNP, but a lower frequency of occurrence (Pitchford et al. 1999).

Following these studies, the EPA issued its final determination that certain visibility impairment episodes at the Grand Canyon were traceable to the NGS, which was a dominant contributor to some of these episodes. Although the contradictory tracer test results emphasized the need for better tracer measurements in future studies, WHITEX demonstrated the potential of tracer techniques for single-source attribution (Pitchford et al. 1999).

Discharge of Mine Tailings into Lake Superior

In 1947 the Reserve Mining Company, then contemplating a venture in which it would mine low-grade iron ore (taconite) in Minnesota's Mesabi iron range, received a permit from the state of Minnesota to discharge its processing wastes into Lake Superior. Taconite mined near Babbitt, Minnesota, was shipped by rail to the Silver Bay beneficiating plant, where it was concentrated into pellets containing 65 percent iron ore. This process involved flushing residual mine tailings, which entered the lake as slurry, then settled on the lake bottom—approximately 67,000 tons of tailings were discharged daily.

Litigation commenced in 1972 when the United States filed a complaint alleging that the company's discharge of tailings violated the Rivers and Harbors Act of 1899, the pre-1972 Federal Water Pollution Control Act, and the federal common law of public nuisance *(Reserve Mining Co. v. Environmental Protection Agency* 1975, *Reserve Mining Company et al. v. United States of America et al.* 1974, *United States of America et al. v. Reserve Mining Company et al.* 1974). The plaintiffs sought an injunction ordering Reserve Mining Company to cease discharging wastes from its processing plant. The Reserve Mining Company maintained that the tailings settled to the bottom of the lake in an area within close range of the plant and that any contaminants present in other areas of Lake Superior could not have originated from Reserve's activities.

To establish their case, the plaintiffs analyzed samples of the lake water or bottom sediment by x-ray diffraction. Upon identifying the element cummingtonite, a principal element in Reserve's discharge, they concluded that the sample contained tailings from Reserve's discharge. Upon quantifying the amount of cummingtonite present in the sample, the plaintiffs were able to determine the quantity of tailings from the discharge. They were also able to show that there were no significant natural sources of cummingtonite in the area and that it was therefore an accurate tracer for the company's discharge.

Based on these results, the court inferred that the actual area of Lake Superior despoiled by the waste from Reserve was more than 2,000 square miles, or an area approximately the size of Delaware. The court found in favor of the plaintiffs by deciding that the discharge represented a state nuisance violation as well as violated the Federal Water Pollution Control Act and the Rivers and Harbors Act of 1899.

Fish Exposure to Pulp and Paper Mill Effluent

In May 1992, the federal government of Canada passed amendments to the Pulp and Paper Effluent Regulations (PPER) under the authority of the Fisheries Act. In addition to the changes in effluent limits, the amended PPER required every mill to conduct an environmental effects monitoring (EEM) program. The objective of the EEM program was to determine if the fish, fish habitat, and use of the fisheries resource were being adequately protected by the requirements of the PPER (Ali et al. 1997).

A tracer program was instituted to assess fish exposure to mill effluent. This information was to be used for investigating biological, physiological, or biochemical indicators of fish health. The use of tracers was based on the premise that chemicals discharged in mill effluent would be present at higher concentrations in fish that frequent these waters than in those from other waters. A total of 100 chemical tracers, including chlorinated dioxins and furans, resin and fatty acids, and chlorinated phenols were monitored in twenty-seven species of fish and crab. Results were reported by seventy-seven mills; some used more than one species, and several different chemical tracers, providing 183 data sets from which the effectiveness of using tracers to verify exposure was assessed (Ali et al. 1997).

Due to a number of technical and operational factors, however, the study did not result in any significant conclusions about effluent exposure. Many of the chemicals that were monitored were below detection, and many of the concentrations measured were close to their detection limits. Other difficulties included apparent inconsistencies among the analytical laboratories on the chemical measurements that were reported. Furthermore, in most cases, only a single composite sample from the reference and study site was used, and little information on chemical concentrations in the effluent was available to compare with information obtained from observed fish. Because of these difficulties, conclusions could not be made regarding the exposure of fish to the pulp and paper effluents, despite the large amount of gathered data. Even if the selected tracers had been successful at determining fish exposure to mill effluents, the unrestricted movement of fish would have precluded further conclusions on the duration of exposure (Ali et al. 1997).

Contamination Distribution

When parameters affecting contaminant transport can be effectively quantified, the distribution of the contamination itself can be used as a basis for determining the source, regardless of the composition of the source effluent. The effectiveness of methods based on this premise depends on the accuracy with which the transport behavior of the specific contaminant, as well as the medium within which it is traveling, can be characterized.

Methods based on the contaminant distribution span a large range of complexity. A common problem of water pollution that illustrates a relatively simple application is the identification of an unknown pollutant source along a river after its detection at a monitoring station (Whitehead, Williams, and Hornberger 1986). Because contaminants disperse and mix as they move downstream, the degree of mixing at a given location can determine the distance traveled by a contaminant. This is based on estimating mixing coefficients (e.g., Peclet numbers, dispersion coefficients) from tracer experiments (Whitehead, Williams, and Hornberger 1986). These methods are more easily applicable to conservative rather than reactive contaminants (Rajagopal 1984).

On the other end of the spectrum are cases of groundwater contamination, in which aquifer pollution is often detected by a fortuitously located single water supply well or group of wells. Transport by groundwater is strongly affected by parameters that are difficult to determine, such as the distribution of hydraulic conductivities and degradation rates in the subsurface. Groundwater models are frequently used to determine sources of observed contamination. Unfortunately, the information on the current distribution of contaminants and hydrogeologic conditions is generally insufficient to allow for a unique solution for the location of sources and the timing of their releases. Despite this limitation, such models can be used to help set bounds on the range of possible contributions from individual sources (National Research Council 1990, 191–200).

Because existing observation wells generally do not have the most efficient locations for identifying the pollution sources in terms of extent, location, and duration of activity (Mahar and Datta 1997), information from additional wells is often necessary. Because of the relatively high cost of installing monitoring wells for contaminant source and extent identification, their deployment is optimized to adequately define the extent of the plume, thus minimizing the number of wells needed. One way to optimize well deployment is to solve an inverse contaminant transport problem, which involves tracking the contamination backward in time to determine its original location (Sidauruk, Cheng, and Ouazar 1998). Such a forensic problem is nonunique and associated with potentially large uncertainty. Interpretation of the estimated results therefore requires careful consideration in the context of other available information. Furthermore, application of inverse methods is highly dependent on the validity of the fate and transport model assumed. Quantitative interpretation is also biased by any variability of the soil properties beyond that reflected in the modeling parameters (Liu and Ball 1999).

A second strategy for source identification is to make a preliminary identification of sources using the observed concentration data from the existing well locations as well as maps, records, and interviews. Then the preliminary identification results may be used to design a more efficient monitoring network. Once such a network is implemented, data subsequently collected from these monitoring wells can be further utilized to more efficiently and accurately identify the pollutant sources (Mahar and Datta 1997).

Examples of methods based on the contaminant distribution are presented for a case in which a single source was examined and for two cases in which several potential sources were present.

Groundwater Contamination from a Single Chemical Waste Burial Site

The plaintiffs in the case of *Woodrow Sterling* (1986) were a class of people who owned property or lived within a three-mile radius of a 242-acre chemical waste burial site in Hardeman County, Tennessee. The site was owned and operated by Velsicol from late 1964 until it was closed as hazardous in 1973 by order of the state of Tennessee. Plaintiffs sought damages for personal injury and damages to their property allegedly suffered when water in their home wells became contaminated by hazardous chemicals that escaped from Velsicol's burial site. The court found Velsicol liable to plaintiffs upon the legal theories of strict liability, common law negligence, trespass, and nuisance. Furthermore, the court found Velsicol's actions in locating, creating, maintaining, and operating its chemical waste burial site constituted gross negligence and a willful and wanton disregard for the health and well-being of the plaintiffs and adjacent environment.

Velsicol admitted, almost from the beginning of the lawsuit, that some of the wells along Toone-Teague Road, used by families as a source of home water supply, were contaminated with chemicals from its waste burial site. Velsicol disputed that all members of the class had been exposed and did not agree with the plaintiffs as to the intensity (concentration or amount) and duration of exposure. Therefore, computer-based modeling of groundwater flow and chemical transport was introduced by plaintiffs and defendants in an attempt to show the intensity and duration of exposure.

Both parties used the same computer models, namely the Trescott, Pinder, and Larson model for flow and the Konikow and Bredehoeft model for transport. Therefore, any differences in the results were caused by the different loading rate factors that were used as inputs and differences in assumed porosity and hydraulic conductivity. Velsicol's model assumed that very little liquid was placed into the dump. For this reason, the court believed that the defendant's estimates were faulty, both as to when the wells became contaminated and the amounts of contamination. A typical set of modeling data is presented in table 7.1 (note that the approximate well concentration was known in 1978, and the results of both parties reproduced the observed values for that year).

The key point of this case study is that the uncertainty associated with modeling results can appear large, even in cases in which a single set of models is used, reputable modelers are hired by both parties, the exposure intensity is approximately known at at least one time, and there is only one source to be examined.

Groundwater Contamination from an Unknown Source

Toxic solvent contamination of the drinking water wells in Woburn, Massachusetts, received great attention in the cases of *Anne Anderson et al. v.*

Table 7.1. Modeling Results for Carbon Tetrachloride Concentrations in the Sterling and Johnson Wells

Year	*Plaintiff (ppb)*	*Defendant (ppb)*
1969	2	
1970	4	
1971	14	
1972	91	
1973	440	
1974	1,200	9
1975	2,700	100
1976	4,700	640
1977	7,000	3,000
1978	9,400	9,600

Source: Woodrow Sterling et al. v. Velsicol Chemical Corporation (1986).

Cryovac, Inc. et al. (1983 and 1988) and *Anne Anderson et al. v. W. R. Grace & Co. et al.* (1986). These cases involved a leukemia cluster that was allegedly the result of industrial contamination. The ensuing litigation was recorded in the popular novel *A Civil Action* (Harr 1995) and in a Hollywood movie bearing the same title.

In 1964 the city of Woburn installed municipal water supply wells designated G and H in the Aberjona River valley. In 1979 public health officials discovered that the wells were contaminated by toxic solvents. The wells were closed, and the site was eventually listed on the National Priorities List as a Superfund site (Myette, Olimpio, and Johnson 1987). After much investigation, officials of the EPA zeroed in on several potential sources of contamination: a manufacturing plant owned by W. R. Grace & Company, premises controlled by Unifirst Company, and a fifteen-acre parcel of vacant wetland owned by the John J. Riley Company, then a subsidiary of the Beatrice Foods Company.

The plaintiffs all lived near wells G and H during and prior to 1979. In 1982 they sued Grace, Beatrice, and others, alleging that the solvents in the water supply had caused them to contract a variety of ailments, including leukemia. Their action was based upon the common law tort theories of negligence, nuisance, and strict liability. The district court opted to try the matter in three stages. In the first phase, the jury would determine whether the defendants were responsible for polluting the wells. The second and third phases, if needed, would be devoted to causation and damages, respectively.

Hydrogeologic modeling was insufficient to conclusively determine the source of the contamination at the time of the trial, mostly due to gross discrepancies between results presented by the plaintiffs' and defendants' expert witnesses. Specifically, the jury found that the results of the modeling studies did not provide sufficient evidence that any of the contamination had originated from the fifteen acres owned by Beatrice, and the case against Beatrice was therefore dismissed. The jury found Grace liable for contaminating the wells with trichloroethylene and tetrachloroethylene. By 1986, when the second-phase trial against Grace was to start, the plaintiffs had settled with W. R. Grace (for an amount that provided only minimal compensation to the plaintiffs), due in large part to insufficient funds to continue the litigation. Further investigations by the EPA eventually showed that the contamination most likely originated from the John J. Riley Tannery (Harr 1995, 451).

Groundwater Contamination from Multiple Sources

Groundwater within a zone approximately six miles long and one mile wide in the vicinity of the Tucson, Arizona, airport is contaminated with organic solvents, primarily trichloroethylene (TCE). The contaminated groundwater is part of an extensive alluvial aquifer that is used as a major source of drinking water in the Tucson area. This is one of the largest metropolitan areas in the country that is almost totally dependent on groundwater for drinking water, and the TCE contamination was viewed as a threat to the water supply. The area containing the contaminated groundwater is listed on the National Priorities List and is known as the Tucson Airport Area Superfund Site. The EPA identified several potential sources of groundwater contamination in a remedial investigation conducted at the site, but there were no data to indicate how much TCE had been released to the subsurface at any individual facility (National Research Council 1990, 191–200).

The investigators arrived at some general conclusions regarding the significance of individual source areas, but these results were questioned by the potentially responsible parties. As a result, the EPA asked CH2M Hill, an environmental consulting firm, to conduct an assessment of potential sources. The areas contaminated by individual sources were estimated using a two-dimensional numerical contaminant transport model, but the estimates of uncertainty about the releases were subjective owing to lack of data. Therefore, only a qualitative assessment was made of the range of relative contributions. The modeling study conducted by CH2M Hill for the EPA was severely criticized by several of the potentially responsible parties, and as a result no agreement was reached on an appropriate remedial action (National Research Council 1990, 191–200).

These shortcomings once again illustrate some of the limitations associated with applying modeling results in a legal setting.

APPLICATION FEASIBILITY FROM A SCIENTIFIC PERSPECTIVE

Based on the discussion of the three major types of methods available for contaminant source identification, it is evident that each type involves certain unique advantages but also has unavoidable limitations. A comparison based on scientific criteria is presented to examine these differences.

The methods differ in terms of their range of applicability to various contaminants. In theory, the distribution of any contaminant can be analyzed or modeled, and the methods based on contaminant distributions are therefore universally applicable, provided the transport can be characterized. The use of tracers, however, whether they be naturally occurring or introduced, depends on a tracer's ability to reproduce the transport behavior of a given contaminant. Such behavior reproduction is not always possible, especially in complex media such as heterogeneous aquifers. Compositional analysis can be applied only in cases in which the contaminant exhibits identifiable traits that are unique to a given source, and its applicability is therefore the most limited.

The number of sources that can be distinguished using the three methods varies in a similar manner. With modeling tools it is possible to determine the contaminant distribution resulting from any number of distinct sources. The number of sources that can be discerned using tracers is limited, at any one time, to the amount of distinct acceptable tracers found in individual potential sources. Compositional analysis offers the strongest limitations, because the number of sources that can be distinguished depends on the number of identifiable contaminant components and their correlations in various sources.

Of the three methods, tracers offer some limitations to reuse. Because a given tracer can accumulate in an environment through sorption, deposition, or other processes, a single tracer can sometimes be reused a limited number of times. Compositional analysis and contaminant distribution analysis, however, do not affect the environment in which the contaminant is being transported, and can therefore be applied infinitely.

Tracers can also be limiting in terms of the time scales required for the analysis. Specifically, if an introduced tracer is used, it must travel to a given receptor before any conclusions can be reached. Compositional analysis and methods based on contaminant distributions do not present such limitations.

Another disadvantage of the use of tracers is that in an overwhelming majority of cases the tracer concentration is lower than that of the contaminant, and dilution effects can result in the tracer being below or very close to detection limits, making source identification impossible. With introduced tracers, the amount of tracer required has to be very high to offset the effects of dilution (Ali et al. 1997). Once again, compositional analysis and methods based on the contaminant distribution do not have similar limitations. If a contaminant is present at detectable levels, then its individual components will most likely also be detectable, and analysis can be made. Similarly, when the contaminant distribution is being examined, the contaminant is clearly above its detection limit if it is of concern.

The three methods differ in their ability to reproduce the behavior of a contaminant. A tracer is effective only if it is transported in a manner similar to the contaminant it is tracing. Erroneous conclusions can result if the tracer has transport characteristics (e.g., sorption, deposition, dispersion) different from those of the contaminant. Determining the source based on the contaminant distribution is even more prone to such errors: Not only do the transport characteristics of a given contaminant need to be understood, but also the characteristics of the medium in which the contaminant has traveled need to be determined. This issue is especially important when contaminant transport modeling is attempted. The intrinsic variability of a given medium can greatly affect transport properties, and such variability must be incorporated into the model being used. Compositional analysis does not suffer from such limitations, because the contaminant is used as a tool for source determination.

Table 7.2 provides a summary of the applicability of the three examined types of contaminant source identification methods based on the discussed scientific criteria. Based on the above discussion, it appears that the optimal

Table 7.2. Comparison of Contaminant Source Identification Methods from a Scientific Perspective

Criteria	*Compositional Analysis*	*Tracer Use*	*Contaminant Distribution*
Range of applicability	Limited	Fair	Good
Number of sources	Limited	Fair	Good
Possibility for reuse	Good	Fair	Good
Time scales required	Good	Fair	Good
Dilution effects	Good	Limited	Good
Behavior reproduction	Good	Fair	Limited

methods vary greatly with application. Furthermore, the advantages and limitations inherent to the three types of methods suggest that using a combination of methods may be the most feasible alternative. A large number of methods are available that often allow reliable source identification.

APPLICATION FEASIBILITY IN CURRENT POLICY SETTING

Although the relative advantages and disadvantages of the three major types of methods of contaminant source identification have been presented from a scientific standpoint, this does not completely determine the effectiveness of the methods in cases in which a legal decision has to be made based on analytical or experimental results. To determine whether the methods are applicable, a number of other criteria need to be examined.

If the presented conclusions are to be convincing to people unfamiliar with the scientific details of the methods, the basic principles underlying the methods must be relatively simple. For example, the concept behind compositional analysis is straightforward. If the contaminants resulting from various sources are measurably different, then conclusions based on these differences are easy to understand. A similar idea applies to the use of tracers. If a tracer is unique to a source and finds its way to a receptor, then the link between source and receptor is easy to make. The concepts underlying the use of the contaminant distribution are not as simple, however. If, for example, the degree of mixing is used as an indicator of the proximity of a source, then two experts can easily suggest two different mixing rates, which would make the probable source appear to be at two different distances upstream of the contaminant location. When contaminant transport modeling is used, many decisions that may appear subjective need to be made, making the results difficult to judge by an untrained audience.

The concept of conceptual simplicity is closely related to that of the weight of the evidence provided by the various methods. If the method is based on assumptions that appear subjective to a lay audience, the weight of the evidence is lesser and the conclusions are much more likely to be contested. Therefore, as with the conceptual simplicity criterion, methods based on contaminant distributions are less attractive relative to compositional analysis and tracer-based methods.

Another factor that must be taken into account is the cost of source identification. Extensive monitoring is required for tracer methods, which can represent an unreasonable expense, especially if long-term studies are required. For methods based on the contamination distribution, the transport processes must be understood, which generally requires substantial characterization of transport parameters. Such characterization can represent a large fraction of the eventual cost of remediation. Methods based on the contaminant composition may require expensive laboratory analyses of contaminant samples, but the overall cost can be lower than that of the other methods, because an extensive monitoring network is usually not necessary.

The current applicability of all discussed methods is somewhat limited given present environmental policies. Nearly all of the federal environmental policies, with the exception of the Comprehensive Environmental Response, Compensation and Liability Act of 1980 (CERCLA) and the Oil Pollution Act of 1990, constitute permitting schemes, as opposed to liability schemes (Marzulla and Marzulla 1997, 93–109). As such, these policies ensure that discharges are within prescribed limits, but do not offer a legal basis for individual property owners who are affected by contamination to defend their rights, unless the contamination is a result of negligence.

Such regulations effectively limit the property rights of both the polluter and the exposed individuals. In such a context, contaminant source identification can be used only as a tool for devising reasonable limits on allowable discharges. These limitations explain the relatively small number of cases in which scientific methods of source identification have been applied in a legal setting. But CERCLA imposes joint and severe strict liability for on-site contamination upon the owner, operator, transporter, or arranger for disposal of hazardous substances. For contamination of adjacent property, it requires proven causation or evidence that the contamination originates on the property of the defendant. Such regulation offers a much broader forum for the application of scientific methods of contaminant source identification because the polluter is liable for the effect of the contaminant release. In this light, current regulations are equally restrictive to all methods of source identification.

Table 7.3 provides a summary of the applicability of the three examined types of contaminant source identification methods based on policy and social criteria. In short, current policies tend to focus on remediation and control of point-source discharges, rather than on source identification. The weight of the evidence is reduced by the fact that some source identification

Table 7.3. Comparison of Contaminant Source Identification Methods from a Policy Perspective

Criteria	*Compositional Analysis*	*Tracer Use*	*Contaminant Distribution*
Conceptual simplicity	Good	Good	Limited
Weight of evidence	Good	Good	Limited
Method cost	Fair	Limited	Limited
Current applicability	Fair	Fair	Fair

methods are based on complex scientific principles, which results in certain variables appearing subjective to a lay audience. The cost of the analyses required for all methods of source identification is substantial.

POTENTIAL FOR FAVORABLE POLICY MODIFICATIONS

The analysis presented in the previous section suggests that policy modifications need to take place before contaminant source identification can become an intrinsic part of the environmental regulatory process. American environmental policy has moved an extraordinary distance over its history, from laissez-faire exploitation of natural resources and dilution as the solution for pollution to active public management and transformation of the environment and, more recently, to detailed national regulation of human impacts. The reality of U.S. environmental policies at the start of the twenty-first century is that the dominant policy approach of the modern environmental era—federal standard-setting, permitting, and enforcement—is no longer sufficient to achieve significant further progress, and that other policies remain fragmented, inconsistent, and often counterproductive (Andrews 1999, 366–72).

Recent shifts in environmental policy ideas, however, could help with the application of source identification technologies. For example, in 1997 a broad working group of environmental policy experts proposed a next generation of environmental policies intended to move the debate over American environmental policy beyond its current stalemate. In the industrial sector, for instance, it was suggested that manufacturing could be more efficiently organized along the principles of clean production and industrial ecology. If industries were to be held responsible for materials and energy throughout their life cycles of human extraction and use, they might find it efficient to redesign their processes to maximize environmental efficiency. A second promising development is the growing use of self-implementing policies. Such policies include information disclosure requirements that compel reporting of data on businesses' environmental performance, and liability policies that hold businesses responsible for the full environmental costs and consequences of their products and processes (Andrews 1999, 366–72). Both these shifts represent increased recognition of the property rights associated with activities resulting in contamination.

There are also economic reasons for the use of contaminant source identification. First, the amount of public funds allocated to cleanup is greatly reduced once a polluter has been identified. Furthermore, the overall cost of remediation can be reduced if the party responsible for the contamination can negotiate an optimal level of pollution with receptors, as opposed to being forced to comply with stringent policy-mandated contaminant discharge levels. Identifying and enforcing the property rights of the receptors and polluters allows market mechanisms to identify economically optimal solutions to given problems.

The fundamental shifts in ideas regarding environmental policy, combined with the economic incentives associated with contaminant source identification, may accelerate the process of implementation of policies incorporating the use of source identification technologies. In order for a system of property rights to be effective at managing contamination, however, the transaction costs associated with such a system must be minimized. The ease and cost of contaminant source identification vary greatly, both with the nature of the contaminant and of the medium. For some types of contamination, current technology is sufficient to identify probable sources. For contaminants and media that are more difficult to characterize, however, reliable source identification is not always possible. Contaminant source identification can help define and enforce the property rights of all parties affected by contamination, but not without some costs. Additional scientific research is needed to develop reliable cost-effective source identification methods that will reduce the transaction costs associated with this property rights approach.

REFERENCES

Ali, N., I. Cloutier, S. Humphrey, A. Niimi, R. Parker, C. Wood, and V. Zitko. 1997. *Exposed or Not Exposed: The Effectiveness of Using Tracers to Verify Exposure of Fish to Pulp and Paper Mill Effluent.* Burlington, Ont.: Department of Fisheries and Oceans.

Anderson, Terry L., and Donald R. Leal. 1991. *Free Market Environmentalism.* San Francisco: Pacific Research Institute for Public Policy and Westview Press.

Andrews, Richard N. L. 1999. *Managing the Environment, Managing Ourselves: A History of American Environmental Policy.* New Haven, Conn.: Yale University Press.

Bentz, Alan P. 1978. Who Spilled the Oil: Matching an Oil Spill with Its Source Requires Methods that Take into Account the Weathering of Oil in the Spill. *Analytical Chemistry* 50: 655A–58A.

Bergamaschi, P., C. A. M. Brenninkmeijer, M. Hahn, T. Roeckmann, D. H. Scharffe, P. J. Crutzen, N. F. Elansky, I. B. Belikov, N. B. A. Trivett, and D. E. J. Worthy. 1998. Isotope Analysis Based Source Identification for Atmospheric CH_4 and CO Sampled Across Russia Using the Trans-Siberian Railroad. *Journal of Geophysical Research* 103(D7): 8227–35.

Brown, Leslie, James Braven, Michael M. Rhead, Roger Evens, and E. I. Butler. 1982. The Use of Uric Acid as a Method of Tracing Domestic Sewage Discharges in Riverine, Estuarine and Coastal Water Environments. *Water Research* 16: 1409–11.

Cable, Jaye E., Glynnis C. Bugna, William C. Burnett, and Jeffrey P. Chanton. 1996. Application of ^{222}Rn and CH_4 for Assessment of Groundwater Discharge to the Coastal Ocean. *Limnology and Oceanography* 41: 1347–53.

Coase, Ronald H. 1960. The Problem of Social Cost. *Journal of Law & Economics* 3(1): 1–44.

Crawford, Mark. 1990. Scientists Battle over Grand Canyon Pollution. *Science* 247: 911–12.

Harr, Jonathan. 1995. *A Civil Action.* New York: Random House.

Horwitz, Robert B. 1988. *The Irony of Regulatory Reform.* New York: Oxford University Press.

Kucklick, J. R., and T. F. Bidleman. 1994. Organic Contaminants in Winyah Bay, South Carolina. II: Using Natural Fluorescence to Follow Atrazine Levels and River Mixing. *Marine Environmental Research* 37: 79–91.

Liu, Chongxuan, and William P. Ball. 1999. Application of Inverse Methods to Contaminant Source Identification from Aquitard Diffusion Profiles at Dover AFB, Delaware. *Water Resources Research* 35: 1975–85.

Mahar, Pooran S., and Bithin Datta. 1997. Optimal Monitoring Network and Ground-Water-Pollution Source Identification. *Journal of Water Resources Planning and Management* 123: 199–207.

Mansuy, Laurence, R. Paul Philp, and Jon Allen. 1997. Source Identification of Oil Spills Based on the Isotopic Composition of Individual Components in Weathered Oil Samples. *Environmental Science and Technology* 31: 3417–25.

Marzulla, Nancie G., and Roger J. Marzulla. 1997. *Property Rights: Understanding Government Takings and Environmental Regulation.* Rockville, Md.: Government Institutes.

Myette, Charles F., Julio C. Olimpio, and David G. Johnson. 1987. *Area of Influence and Zone of Contribution to Superfund-Site Wells G and H, Woburn, Massachusetts.* Boston: U.S. Geological Survey.

National Research Council. 1990. *Ground Water Models: Scientific and Regulatory Applications.* Washington, D.C.: National Academy Press.

Pitchford, Marc, Mark Green, Ival Tombach, William Malm, and Robert Farber. 1999. *Project Mohave Final Report.* Washington, D.C.: U.S. Environmental Protection Agency.

Rabinowitz, Michael B. 1995. Stable Isotopes of Lead for Source Identification. *Clinical Toxicology* 33: 649–55.

Rabinowitz, Michael B., and George W. Wetherill. 1972. Identifying Sources of Lead by Stable Isotope Techniques. *Environmental Science and Technology* 6: 705–709.

Rabinowitz, M., G. Wetherill, and J. Kopple. 1977. Magnitude of Lead Intake from Respiration by Normal Men. *Journal of Laboratory and Clinical Medicine* 90: 238–48.

Rachdawong, P., E. R. Christensen, and S. Chi. 1998. Source Identification of PCBs in Sediments from the Milwaukee Harbor Estuary, USA. *Water Science and Technology* 37: 199–206.

Rajagopal, R. 1984. Optimal Sampling Strategies for Source Identification in Environmental Episodes. *Environmental Monitoring and Assessment* 4: 1–14.

Rusin, Patricia A., Norval A. Sinclair, Charles P. Gerba, and Melvin Gershman. 1992. Application of Phage Typing to the Identification of Sources of Groundwater Contamination. *Journal of Contaminant Hydrology* 11: 173–88.

Sidauruk, P., Ah-D. Cheng, and D. Ouazar. 1998. Ground Water Contaminant Source and Transport Parameter Identification by Correlation Coefficient Optimization. *Ground Water* 36: 208–14.

Stroup, Richard L. 1990. Chemophobia and Activist Environmental Antidotes: Is the Cure More Deadly than the Disease? In *Economics and the Environment: A Reconciliation*, ed. Walter Block. Vancouver, B.C.: Fraser Institute, 193–213.

Surkyn, P., J. de Waele, and F. Adams. 1983. Laser Microprobe Mass Analysis for Source Identification of Air Particulate Matter. *International Journal of Environmental Analytical Chemistry* 13: 257–74.

Whitehead, P. G., R. J. Williams, and G. M. Hornberger. 1986. On the Identification of Pollutant or Tracer Sources Using Dispersion Theory. *Journal of Hydrology* 84: 273–86.

Zweidinger, Roy B., Robert K. Stevens, and Charles W. Lewis. 1990. Identification of Volatile Hydrocarbons as Mobile Source Tracers for Fine-Particulate Organics. *Environmental Science and Technology* 24: 538–42.

CASES CITED

Anne Anderson et al. v. Cryovac, Inc. et al., 96 F.R.D.; 1983 U.S. Dist. LEXIS 19830; 37 Fed. R. Serv. 2d (Callaghan) 980; 19 ERC (BNA) 1405 (1983).

Anne Anderson et al. v. Cryovac, Inc. et al., 862 F.2d 910; 1988 U.S. App. LEXIS 16574; 12 Fed. R. Serv. 3d (Callaghan) 719; 28 ERC (BNA) 1622 (1988).

Anne Anderson et al. v. W. R. Grace & Co. et al., 628 F. Supp. 1219; 1986 U.S. Dist. LEXIS 30863; 23 ERC (BNA) 2070; 16 ELR 20577 (1986).

Central Arizona Water Conservation District et al. v. United States Environmental Protection Agency, 990 F.2d 1531; 1993 U.S. App. LEXIS 5881; 36 ERC (BNA) 1177; 93 Cal. Daily Op. Service 2159; 93 Daily Journal DAR 3793; 23 ELR 20678 (1993).

Ethyl Corporation et al. v. Environmental Protection Agency, 176 U.S. App. D.C. 373; 541 F.2d 1; 1976 U.S. App. LEXIS 12309; 8 ERC (BNA) 1785; 6 ELR 20267 (1976).

Lopardo v. Fleming Cos., 97 F.3d 921; 1996 U.S. App. LEXIS 26191 (1996).

Reserve Mining Company et al. v. United States of America et al., 498 F.2d 1073; 1974 U.S. App. LEXIS 8322; 6 ERC (BNA) 1609; 4 ELR 20598 (1974).

Reserve Mining Co. v. Environmental Protection Agency, 514 F.2d 492; 1975 U.S. App. LEXIS 15664 (1975).

United States of America et al. v. Reserve Mining Company et al., 380 F. Supp. 11; 1974 U.S. Dist. LEXIS 8891; 6 ERC (BNA) 1449; 4 ELR 20573 (1974).

Woodrow Sterling et al. v. Velsicol Chemical Corporation, 647 F. Supp. 303; 1986 U.S. Dist. LEXIS 21973; 24 ERC (BNA) 2017; 17 ELR 20081 (1986).

8

Property Rights and Technological Innovation: Legal Remedies and Pollution Abatement in U.S. Mining

David Gerard and Timothy J. LeCain

A host of environmental problems arise when the actions of one party result in ongoing harm to a neighboring party. In such cases, the affected party can seek relief through traditional common law actions of nuisance and tort. If the court finds for the plaintiff, the legal remedies available are injunctive relief (a property rule) and damage awards (a liability rule).[1]

Despite the prominent role of nuisance law in dealing with environmental problems, there has been no attempt to examine whether the choice of a legal remedy affects the design and implementation of new abatement technologies. This chapter addresses the issue in the context of U.S. hardrock mining industry of the late nineteenth and early twentieth centuries. Changes in mining and mineral processing technologies during this period spawned phenomenal expansion of mineral output, but also contributed to escalating levels of pollution. In particular, hydraulic mining operations in California and copper and lead smelters across the United States caused major environmental disruptions (Kelley 1959; LeCain 1998).

There is little mystery why firms searched for new mining and mineral processing technologies. The promise of fantastic wealth, coupled with an accommodating institutional structure, provided the incentive for firms to find new ways to separate the valuable metals from worthless rock.[2] The incentives to develop and implement new pollution-abatement technologies, however, were less obvious. Assuming profit maximization, firms will search for ways to reduce pollution as long as the benefits of doing so exceed the costs. Not surprisingly, firms often saw few, if any, benefits associated with mitigating offsite impacts of their operations, and consequently were reluctant to design and implement pollution-abatement measures if there was no clear benefit of doing so.

Parties damaged by the offsite impacts of mining and mineral processing operations often brought suit under traditional common law actions of nuisance and tort. The legal remedies were injunctive relief, compensatory damages, or a combination of the two. A basic result of the law and economics literature is that when costs of enforcing contracts are low and parties have full information, either legal remedy will elicit the efficient result (Polinsky 1980b, 1081).

But in this chapter we find that injunctive relief, or the specter of injunctive relief, was a primary determinant of the adoption of pollution-abatement technologies at copper and lead smelters. With few exceptions, the first firms to adopt pollution-abatement measures were responding to court orders to mitigate offsite damages. Several smelter operations shut down or relocated in response to these injunctions, whereas others reduced output or implemented new pollution-abatement technologies.

During this period, U.S. courts became more reluctant to grant injunctive relief as a remedy for a nuisance action when the harm to the defendant (and the community at large) was deemed to be greater than the benefit that would accrue to the plaintiff if the injunction was granted (Brenner 1974; Horwitz 1977; Rosen 1993). In this decision process, known as a balancing test, the courts balanced relative hardships of the litigating parties rather than protecting the entitlements of private property rights. Thus, courts that conducted balancing tests were more likely to award damages to the plaintiff, but let the nuisance continue unabated (Rosen 1993). Few, if any, new pollution-abatement technologies were introduced following these cases.

This is a result in conflict with Posner (1992), who argues that the common law develops efficient rules, and therefore the emergence of a balancing test reflects an efficient innovation.[3,4] Although Posner (1992, 63–65) acknowledges that this can affect incentives to adopt abatement measures, he does not speculate on whether it retards the search for new technologies. Given that the introduction of new pollution-abatement technologies generally followed a court injunction, however, it is possible that the more frequent use of balancing tests slowed the rate at which new pollution-abatement technologies were designed and implemented.[5]

This presents an interesting conundrum. Even if the court's balancing analysis allocates the right to the highest-value user in the short run (e.g., in the case of a damage award to the plaintiff from a defendant), such decisions might retard the design and implementation of effective pollution-abatement technologies in the long run. In the rest of this chapter, we develop a hypothesis and present evidence as to why this can occur.

In the following two sections, we review the development of U.S. mining industry following the California gold rush, including the adoption of major pollution-abatement technologies at copper and lead smelters. The third section develops the basic distinctions and economic implications of property and liability rules, and is followed by a description of the results of nuisance actions against hydraulic mine operators in California and copper and lead smelters. The last section offers concluding comments.

CAPITAL, TECHNOLOGY, AND THE RISE OF KING COPPER

By the eve of the California gold rush in 1848, metal mining technology had advanced little beyond what existed during the medieval period. U.S. mining was relatively rudimentary and primarily focused on precious metals development. Small gold or silver mining companies operated using hand labor and basic technology; the valuable and compact output was transported by pack trains and wagons. Small-scale ventures were often able to extract very high-grade ore from a new site, but handling lower-grade material generally required large capital outlays and technical expertise—scarce assets for small enterprises. In one way or another, a site initially worked by a number of producers usually was consolidated by a single owner. As rich surface deposits played out, large firms began to dominate precious metals production.

California and Nevada saw the beginning of a rapid series of important technological developments in American mining. By the mid-1850s in California, for instance, capital-intensive hydraulic operations were reworking areas abandoned by small-scale placer miners (Swenson 1968, 710). The *Mining and Scientific Press* called California hydraulic operators "among the greatest marvels of engineering skill the world can boast of" (Holliday 1999, 275). Large operations eventually assumed control of mining operations on Nevada's Comstock Lode and received similar accolades for technical achievement. Square-set timbering, for instance, demonstrated the viability of deep-level metal mining, and the practice was first successfully applied in the United States on the Comstock. Such extensive capital requirements "ended the poor man's day in mining, and ushered in the era of the financier and engineer" (Barger and Schurr 1944, 101).

Even the Comstock operations were small compared with those needed for copper mining, where access to capital was considered a prerequisite for success (Navin 1978). Most copper projects could not be pursued successfully on a small scale; they demanded expensive rock drills or steam shovels, extensive rail lines for haulage, massive milling and smelting operations, and transportation infrastructure to ship off the low-value output.[6] The change of the industry focus toward the large, low-grade deposits of copper, lead, zinc, molybdenum, and other base metals required major capital outlays. Consequently, capital-intensive and technologically sophisticated firms came to dominate western mineral production. The three largest western copper companies—Kennecott, Phelps Dodge, and Anaconda—began as capital-intensive operations with large investments in advanced mining technology and engineering expertise (Navin 1978, 6). By the turn of the century, many of the largest precious metals mines were copper and lead mines, with silver and gold captured as valuable secondary minerals during the production process.

The results were phenomenal. Between 1880 and 1910, the annual output of gold from mines in the United States increased from about 1.5 million ounces to more than 4.5 million ounces. During the same period, American copper production expanded more than eighteen times, and the production of other base metals soared as well. The total value of metals produced in 1880 was $125 million; thirty years later the figure had nearly quadrupled to $470 million.[7]

POLLUTION AT COPPER AND LEAD SMELTERS

The advance and dispersion of technology was central to the economic ascent of the American mining industry. Although technology was widely accessible, the practices often had to be tailored to the characteristics of a given site. Different mines had all sorts of different technological demands, from what type of dynamite was most effective to the best milling machines for their ore. In terms of pollution, the smoke resulting from the smelting of any particular ore is often unique, and, in many cases, a pollution-control technology that works for one smelter fails for another. Smoke pollution controls, like production technologies, often had to be customized to conditions at each smelter.

The generation and distribution of technical information were facilitated by a number of factors. An expanding body of professional mining engineers and other professional associations, such as the American Institute of Mining Engineers, commercial publications, including the *Engineering and Mining Journal* and the *Mining and Scientific Press*, were all instrumental in disseminating information about technological advances and innovations. In addition, the U.S. Bureau of Mines distributed information about new technologies and did its own research and development work.

Western copper deposits were primarily low-grade deposits known as sulfides. These sulfides typically consisted of 1 to 3 percent copper, 25 to 40 percent sulfur, and a host of other impurities like calcium, iron, and arsenic. After the ore was mined, the smelting process separated the metals from the rest of this material. This was done by heating the rock to an extremely high temperature to liquefy the ore. During this process, the heavy metals settled below the worthless molten rock, creating considerable waste materials. The molten rock floating on the surface, known as slag, was skimmed off and disposed of as solid waste. The smelting process also drove substantial material out of the ore, discharging particulate matter (including heavy metals), fumes and vapors, and gases into the air. As a result of the smelting process, sulfur and arsenic combined with oxygen to form sulfur dioxide, arsenic trioxide, and other noxious gases and particles.

Early smelting operators made no attempt to curb the release of the arsenic and the sulfuric gases, and consequently smelter regions were plagued with billowing clouds of foul-smelling, brownish yellow smoke. Even relatively small and isolated early smelters produced enough smoke to create serious problems. In the 1870s, a correspondent from the *Engineering and Mining Journal* en route to a Colorado smelter recorded that passengers in his train had to endure "villainous vapors, choking gases . . . and cough-compelling odors" while the train was still more than a quarter of a mile from the smelter (Fells 1979, 49–50).

The environmental and health impacts of releasing arsenical and sulfuric compounds into the atmosphere had always been serious. With the expansion of size and capacity of copper mining and smelting enterprises, the amount and concentration of these pollutants led to more frequent and more serious consequences for plant and animal life. Although it is well known that arsenic is poisonous, it is difficult to demonstrate any human deaths resulting from arsenic poisoning around the copper smelters. Modern knowledge suggests that illness and severe discomfort must have been common, and it is highly probable that arsenic pollution killed thousands of cows, horses, and sheep (Franzle 1993, 224–27).

Sulfur dioxide is also toxic to humans, but it is not generally deadly in the concentrations typically released from copper smelters. The effect on plants and trees at these concentrations, however, could be quite severe. Even small amounts can result in a general lack of health and vitality in many crops, trees, and other plants, and high concentrations can rapidly kill certain crops. Secondary effects thus include long-term damage to farm soils and accelerated soil erosion (Hellenbeck and Cunningham-Burns 1985; Tver 1981).

Abatement Technologies

A number of new or modified abatement technologies emerged in response to these problems: Among the earliest was the tall smokestack. These smokestacks had been used for centuries to release annoying smoke from coal or wood fires away from the ground where people lived and worked. Copper smelting engineers designed tall smokestacks to reduce arsenic and sulfur concentrations. Smelter engineers built ever higher stacks, so that by the time the smoke reached people, animals, and crops on the ground, they argued it would be too diluted to cause any harm. At its most extreme, the Anaconda company built the world's tallest smokestack in 1903, a 585-foot giant that is still the largest such structure in the world (Fulton 1915). In principle, this was not a bad strategy, but in practice, farmers in areas surrounding Anaconda, Ducktown, and Selby argued that the solution simply did not work. Instead, pollution was more widely distributed, often spreading the harmful effects over a larger area.

The Tennessee Copper Company developed a more effective solution to sulfur dioxide pollution. Engineers invented a method of cycling the smoke from their copper smelter through a large recovery plant that removed much, though not all, of the sulfur dioxide and converted it into sulfuric acid. The company was then able to sell the sulfuric acid to other nearby companies, which used it for making a type of high-phosphate fertilizer. In later decades, other uses for sulfuric acid were also developed, making it one of the most important of modern industrial chemicals and creating additional markets for the copper smelters (Fulton 1915, 83–84; Quinn 1993, 575–612; Swain 1949, 2384–87).[8]

Another reasonably effective abatement technology widely adopted by smelter operators was the bag house, which was introduced and refined between 1905 and 1915. As the name implies, the bag house was a large cubical houselike structure where hundreds of bags made of wool and other materials were hung from the ceiling. Engineers routed the smoke stream through this structure so it was forced to pass through multiple layers of bags, capturing some of the particulate pollutants in the fine weave of the fabric. The bags were periodically shaken out, releasing valuable tiny particles of copper, gold, silver, and lead, as well as a certain amount of arsenic trioxide. Not surprisingly, bag houses were a popular pollution-abatement technology because they were able to capture metals that would have been lost. The bag house had a positive environmental impact because these heavy metals were not released and did not contribute to heavy metal contamination of the land surrounding the smelter.

Bag houses also succeeded in removing at least some of the arsenical pollution in smelter smoke, but not all of it because many of the smaller particulates passed right through the fabric. Bag houses were rarely used at copper smelters, however, because the smoke streams were so acidic that they quickly destroyed the expensive woolen bags; consequently, the costs of replacing a whole set of bags every few weeks made this technology prohibitively expensive (Fulton 1915, 49; Peters 1911, 578–97).

From 1905 to 1907, Frederick Cottrell, a Berkeley physical chemistry professor, invented and perfected a more practical abatement technology—the electrostatic precipitator (ESP). The technology was completely operable for copper and lead smelters by 1915. The ESP operates by generating a powerful electrical field in which a steady stream of charged ions moves from an emitter to a collector. When particulates like arsenic trioxide enter the field, they are bombarded by the stream of ions, transferring their charges to the arsenic particles, which are then attracted to and captured by the oppositely charged collecting plate. To be effective, the ESP must be precisely adjusted to the smoke stream conditions at each smelter. In many instances, the ESP was highly successful in capturing the bulk of not only arsenic trioxide and condensed sulfuric acid particles, but also gold, silver, and copper. Usually used in tandem with other abatement technologies like bag houses and sulfuric acid plants, the ESP could substantially reduce the level of arsenic and heavy metal pollution produced by copper smelters (Cameron 1952, 120–24; *Encyclopedia of Chemical Technology* 1977, 677–82).

LEGAL REMEDIES AND TECHNOLOGICAL ADVANCES

A nuisance occurs when the activity of one party causes ongoing harm to a neighboring party. The affected party can seek relief through the courts. For a private nuisance action,[9] the court has two remedies available if it finds for the plaintiff, injunctive relief or compensatory damages. Injunctions are often viewed only as permanent prohibition of an activity, though this is not necessarily the case. An injunction, for instance, can be a stipulation that certain measures be implemented or an assurance that some impact will be controlled. The court, of course, can award damages without enjoining the activity.

The injunction is seen as the establishment of a clear property right that the owners can keep unless they voluntarily choose to part with it, whereas damage awards transfer the property right to the defendant in exchange for monetary compensation determined by the courts. These remedies are known as property rules and liability rules, respectively.

The pioneering work of Calabresi and Melamed (1972) examines when each of these remedies is likely to be the efficient choice for the court. If the absence of clear rights is the impediment to a voluntary exchange and if transaction costs are low (e.g., the costs of reaching and enforcing an agreement), the court should issue an injunction. If, however, even a clear assignment of rights will not lead to a voluntary agreement, then the court should deny injunctive relief and award compensatory damages. In such cases, the court rationale must be that a voluntary transaction was not possible because of high transaction costs.

The argument follows Coasean logic. The injunction is a clear assignment of a property right; once bargaining parties have a clear picture of rights, they are likely to reach a voluntary agreement.[10] With high transaction costs, however, an injunction is unlikely to promote voluntary agreements and it is more efficient for the court to award compensatory damages. This is equivalent to the court forcing plaintiffs to sell rights at a price set by the court.

There are several possible sources of high transaction costs. For instance, there could be a large number of parties with conflicting views on whether there is a harm, the cause of the harm, the value of the damages, and the mechanism for enforcing any agreement. The more parties there are, the less likely voluntary contracting will be effective. Thus, it is common to use the number of bargaining parties as a proxy for the level of transaction costs.

In the context of smelting operations, there were often a number of smelter operations and hundreds of potential plaintiffs, suggesting high transaction costs. Typically, however, a farmers' association would form and act as a plaintiff against the smelters. Thus, a potentially large number of plaintiffs was consolidated as a single bargaining unit, suggesting that a clear assignment of rights through injunctive relief could promote Coasean bargaining. Indeed, in a number of cases, the parties reached mutually agreeable terms after the court issued an injunction. Smelters installed abatement technology to comply with court orders, and many continued profitable operations despite the specific performance orders to reduce their emissions. In other cases, smelters did not try to abate pollution, or they tried and failed, in which case the injunction would be enforced and the smelter would be forced to shut down.

High transaction costs can also result from holdouts, even when there are as few as two bargaining parties. In such cases the owner holds out for compensation far above any reasonable terms (Posner 1992, 61–65). The holdout is inefficient where gains from trade are not realized because one of the bargaining parties acts strategically (e.g., if the defendant's willingness to pay is higher than the plaintiff's willingness to accept, but the plaintiff holds out and an agreement is not reached).[11] Because the property rule provides a stronger bargaining position for the plaintiff than a liability rule, the probability of strategic behavior by the plaintiff increases. As a result, it might be efficient for courts to choose a default rule of damage awards to mitigate the costs associated with holdouts.[12]

Overall, the literature has little to offer in how the choice of a legal remedy affects the design of new technologies. Given that there is no theoretical reason why a compensatory scheme that accounted for full costs of pollution would not induce an optimal level of emissions, it seems that a liability rule would induce the same rate of development of new technologies as a property rule. Our review of the history of mining indicates that this was not the case; injunctions (the property rule) were far more technology enhancing than damages (the liability rule). We offer two reasons for these differences.

First, once the balancing test is introduced and the issue of compensatory damages comes before the court, efficient results occur only if the court has good information about the subjective values of the parties involved. And the court has to award damages that reflect those values. Both requirements for efficiency are problematic.[13]

Second, damages are backward looking rather than forward looking. They represent payment for harm already imposed upon third parties. In contrast, injunctions clarify property rights, and as such are much more adaptable to an unforeseen future. Since much of the technology that reduced mining pollution was either unknown or not perfected at the time of the court action, the property rule favored continued technological developments.

Earlier we discussed the problems that high transaction costs represent for property rules, or injunctions. It is obvious, however, that there are also transaction cost problems inherent in the liability rule, problems of accurately measuring and imposing damages, and of accounting for future events. Evidence indicates that in terms of the evolution of certain technologies in mining, the transaction cost problems were greater for the liability rule.

THE MOVE TO BALANCING TESTS AND COMPARATIVE INJURY

There is a substantial body of work discussing the effects of industrialization on the judicial choice of legal remedies in common law countries throughout the world. The basic pattern is that, prior to industrialization, injunctive relief was the default rule. Horwitz (1977, 74), for instance, describes the situation in the United States: "With almost complete unanimity, American courts before the Civil War continued to echo Blackstone's view that even a lawful use of one's property which caused injury to the land of another could be enjoined as a nuisance, 'for it is incumbent on him to find some other place to do that act, where it will be less offensive.'"[14]

Similarly, in her discussion of Canadian law, Nedelsky (1981, 286) states: "Traditional nuisance law asked only whether the plaintiff's rights were unreasonably interfered with, not how carefully conducted or economically important the defendant's activities were. Strict application of the law should have entitled aggrieved neighbors of smelters and foundries to an injunction."

Brenner (1974) agrees that the default remedy in nuisance cases was injunctive relief, but he identifies few instances in which plaintiffs sought and obtained injunctive relief against industrial polluters after 1850.[15] Instead, he finds that compensatory damages almost immediately displaced injunctive relief as the legal remedy of choice following industrialization in England. Citing an 1866 case, Brenner (1974, 407) argues: "On the whole, injunctions or abatement orders were rare indeed. The general rule was that equity courts would 'not intervene by way of mandatory injunction, except in cases in which extreme, or at all events, very serious damage will arise from its interference being withheld.' Most plaintiffs therefore preferred to seek common law relief in damages."

As part of his explanation underlying the transformation of the application of nuisance law, Brenner discusses a series of cases, beginning in 1858, in which the court balanced the harm to the plaintiff against the importance of the defendant's activities (1974, 411–15). Similarly, as U.S. courts began to balance the interests of the parties involved during the latter half of the nineteenth century, the use of compensatory damages in lieu of injunctive relief as the principal remedy in nuisance cases also emerged. To illustrate, Rosen (1993) looked at nuisance decisions in New York, New Jersey, and Pennsylvania between 1840 and 1906, and found that court decisions that discuss the balancing of interests were much less likely to grant injunctive relief. In contrast, the court granted injunctive relief in 82 percent of the cases in which the court did not balance competing interests. But as balancing became more prevalent, the use of damages increased.

The theory underlying the choice of an efficient legal remedy indicates that in a high transaction cost world, damage awards should simulate results of cooperative bargaining that would have taken place if the barriers to striking an agreement did not exist. One potential source of high transaction costs is the holdout problem. To illustrate, in *Madison v. Ducktown Sulphur, Copper and Iron Co.* (1904), the court did not enjoin operations that were destroying land values in the surrounding area. The court opinion states that the tract of land was worth about $1,000, whereas the Tennessee copper industry produced $12 million annually and employed several thousand workers. One can argue that this was an efficient choice of a legal remedy because high transaction costs would have precluded voluntary contracting after an injunction was granted.

One must remember, however, that the choice of a liability rule here removed control of property rights from the plaintiffs in exchange for damages. But in this and many other smelter cases, allowing the defendant to continue its emission discharges unabated would result in the complete devastation of the surrounding area. Damage awards would have to be very large to account for future activity and it would seem that the court should be cautious about using a liability rule if there is considerable uncertainty about the net present value of the costs and benefits.

A court order to pay damages but that allows pollution to continue is therefore akin to the court brokering an involuntary transaction. Such a solution is efficient and equitable if the value of the damaged property is correctly valued, and the court decision serves to lower transaction costs and results in gains to both parties. It is also possible, however, that the court has imperfect information about the value parties place on an asset. The court recognized this possibility in *ASARCO v. Godfrey* (1907), and reasoned that the absence of a voluntary transaction could well have been the result of discrepancies between valuations by the plaintiff and the defendant:

> The fact that the defendants are willing to pay for the property is immaterial, for there are no means of determining whether the value of the property in money would compensate the plaintiffs for its destruction. It may possess a value to them which no other person would place upon it; and there is neither justice nor equity in refusing to protect them in the enjoyment of it, merely because they may possibly recover what other may deem an equivalent in money. The nature of the property, which consists of fruit trees, ornamental shrubbery, etc., gives them a peculiar claim to this protection. (*ASARCO v. Godfrey* 1907, 233)

Although recent work suggests that damage awards are superior to injunctive relief even where there is considerable judicial uncertainty, this result does not hold if damage awards are not sufficient to cover future costs (Kaplow and Shavell 1996). Dewees finds that courts dramatically underestimated full damages (even absent any possible discrepancies in subjective valuations), implying that compensatory damages were not sufficient to make polluters pay the full costs of their actions. He estimates that the present value of the harm was generally "hundreds or thousands" (1995, 61) of times greater than damage awards—even allowing for imperfect scientific knowledge of property and health damages.[16]

A critical question is why the courts consistently undercompensated plaintiffs. For instance, the courts might have favored dominant economic interests because of direct bribery, political influence, or because firms with deep pockets could afford high-quality defenses. Horwitz (1977) argues that the introduction of the balancing test reflected a broader judicial trend—the systematic tailoring of the law to favor dominant economic interests. Other possibilities are that the courts simply did not grasp the extent of the damages or did not believe testimony of damaged parties.

As the United States and other common law countries industrialized, courts began to employ economic balancing tests when choosing between injunctive relief and compensatory damages as a legal remedy in nuisance cases. In the next two sections, we explore how the application of nuisance law affected the introduction of pollution-abatement technologies in U.S. mining. In some cases, the courts permanently enjoined harmful activities, thus preventing ongoing damages and also prompting firms to try to apply new abatement technologies. In some cases, the resultant technological innovations successfully reduced environmental harms. Still, in other cases, firms tried and failed to apply new abatement technologies, and the courts shut down these activities. In contrast, when compensatory damages were awarded instead of injunctive relief, there was much less observed innovation.

NUISANCE LAW AND HYDRAULIC MINING

The most famous application of nuisance law to U.S. mining involved the tremendous devastation wrought by hydraulic gold mining operations in California. These operations directed high-pressure streams of water at sites with promising concentrations of gold in the soil, often literally washing away mountainsides. The water washed loose rock and gravel through sluices where the gold particles were trapped. The water took the debris—and most of the mountain—downstream, clogging river channels and burying fertile topsoil. When the winter rains came, the debris was often set in motion again as the clogged river channels quickly overflowed, resulting in substantial downstream flooding (Kelley 1959).

Hydraulic mining continued virtually unabated starting in the mid-1850s. As technology advanced, operations began to move even larger quantities of earth. Levees built to protect downstream users failed almost without exception. Moreover, efforts to obtain legislative relief also failed.

As a result, the affected parties turned to the courts for relief. After one California state court overturned an injunction issued by another in 1879, Edwards Woodruff filed suit in federal court in 1882. In *Woodruff v. North Bloomfield Gravel and Mining* (1884, 769), the judge, concurring with counsel, found that these operations constituted "a nuisance destructive, continuous, increasing, and threatening to continue, increase, and still be more destructive." The judge continued: "After an examination of the great questions involved, as careful and thorough as we are capable of giving them, with a painfully anxious appreciation of the responsibilities resting on us, and of the disastrous consequences to the defendants, we can come to no other conclusion than the complainant is entitled to a perpetual injunction" (808–809).

The technical mastery of hydraulic mining was not accompanied by technical advances to prevent devastation of the land. In fact, the primary reason the court granted a perpetual injunction in *Woodruff* was that the prospects for preventing further damage were dim. The judge estimated that 700 million cubic yards of gravel would have been washed away on the Yuba River alone (Holliday 1999, 297–98). Here, the court protections of property rights, however delayed, eventually halted the damages from the offensive activity.

Interestingly, many historians have interpreted this court's decision as based on the comparative economic might of the litigating parties. As agricultural interests became more powerful, historians point out, the balance of power in California shifted away from mining. The shift provided the setting for the court to shut down hydraulic mining operations. The *Woodruff* opinion, however, claims that a balancing test was not applied:

> A great deal has been said about the comparative public importance of the mining interests, and also the great loss and inconvenience to these defendants, if their operations should be stopped by injunction. But these are considerations with which we have nothing to do. We are simply to determine whether the complainant's rights have been infringed, and, if so, afford him such relief as the law entitled him to receive, whatever the consequence or inconvenience to the wrongdoers or to the general public may be. (*Woodruff* 1884, 806)

NUISANCE LAW AND SMELTER CASES

In copper and lead smelter cases, injunctive relief encouraged smelters to find ways to reduce offsite impacts. The discussion in this section includes cases in which the courts awarded damages after conducting a balancing test, cases in which injunctive relief was granted, and cases in which injunctive relief was denied. A summary of litigation and technology adoption is presented in table 8.1.[17]

Balancing

In *Madison v. Ducktown Sulphur, Copper and Iron Co.* (1904, 367) the Supreme Court of Tennessee was clearly reluctant to shut down the copper industry, arguing that the

> defendants cannot reduce their ores in a manner different from that they are now employing, and there is no more remote place to which they can remove. . . . We see no escape from the conclusion in the present case that the only proper decree is to allow the complainants a reference for the ascertainment of damages, and that the injunction must be denied to them.

Cases such as *Madison* provided a clear perception to potential plaintiffs that state courts were unwilling to enjoin mineral processing, and consequently plaintiffs sought standing in federal court. Even so, federal courts also fashioned economic balancing tests in their choice of remedies. Because of the low damage awards, the balancing tests systematically biased their decisions in favor of larger companies.

Consider smelter operations in California. Mountain Copper bought out many area farmers after it began operations, but the U.S. government successfully pursued an injunction for damages to the national forests. On appeal in *Mountain Copper v. United States* (1906), however, the federal court used a balancing test to reverse the injunction order. The court found that the land was "practically worthless for agricultural and horticultural purposes," and that the abatement procedure of acid production was impracticable because of "the enormous cost" and "the impossibility of disposing of such product" (642). In effect, the court denied injunctive relief because of the economic burden it would place on the defendant.

Notably, Mountain Copper was one of the first plants to successfully demonstrate the feasibility of capturing sulfur emissions and producing sulfuric acid. Sulfuric acid is mixed with potash to make fertilizer and can be used to convert ammonia by-product of coking ovens into ammonium sulfate (Fulton 1915, 66). Mountain Copper aborted acid production, however, because the company could not dispose of it profitably. Thus, the court's failure to enjoin the operation reduced the company's (nonpecuniary) incentive to reduce sulfur discharges.[18]

Injunctive Relief and Technological Innovation

In several cases, courts enjoined smelter operations. After failing to find relief in Tennessee state court, the state of Georgia successfully obtained injunctions against Tennessee Copper and Ducktown Copper in federal

Table 8.1. Litigation and Technology Adoption Summary

Plaintiff	*Defendant (State)*	*Year*	*Injunction*	*Key Result*
Madison	Ducktown Sulphur (TN)	1904	No	
United States	Mountain Copper (CA)	1906	No	Smelter discontinues acid production
Godfrey and Farmers' Assn.	ASARCO (UT); Bingham Consolidated (UT); Smelting (UT); Utah Consolidated (UT)	1906	Yes	U.S. Smelting and ASARCO add bag houses at lead smelters; Bingham and Utah relocate copper smelters; U.S. Smelting shuts down copper operations
Georgia	Tennessee Copper (GA and TN)	1907	Yes	First commercially successful acid plant at a copper smelter
California	Selby (CA)	1908	Yes	ESP and bag house installed
Bliss	Anaconda (MT)	1909	No	ESP
Shasta Farmers'	Mammoth Copper (CA) Balaklala Consolidated (CA)	1910	Yes	Mammoth is first copper smelter to install bag house; Balaklala installs an ESP
United States	Anaconda (MT)	1911	Settled	No acid plant installed
Anderson	ASARCO (UT)	1919	Yes	Emissions limited

Sources: Fulton (1915) discusses most of these cases; see case list at the end of this chapter.

court in 1907. The companies were not shut down, however, because they were in the process of installing acid plants to mitigate the offsite impacts. The smelter companies were in the process of implementing acid plants before the court granted injunctive relief, and it is not clear whether litigation prompted them to pursue abatement strategies.

The production of sulfuric acid turned out to be a financial boon for these smelting operations, as acid production soon became more profitable than smelting. This case provides an excellent example of bargaining once an injunction clearly established a property right. The acid plant could not handle the volume of material, and consequently Tennessee Copper agreed to reduce its operations during growing season. The court opinion indicates that the firms reduced their sulfur emissions. Without the plant, 85.5 percent of the sulfur content was released; with the plant, the sulfur content released was reduced to 41.5 percent. But in 1915 Georgia remained unsatisfied with the companies' environmental performance and sought and obtained further restrictions on the smelter operations (*Georgia v. Tennessee Copper Co. and Ducktown Sulphur, Copper and Iron Company, Limited* 1915).

Litigation in Utah also had pronounced effects on the behavior of smelter operators. In 1904, after five years of ongoing damage, a group of 409 farmers filed for an injunction in federal court against a number of smelters in the Salt Lake City area. The federal court enjoined the smelters in 1906, and the appeals court affirmed the injunction a year later. The *Godfrey* case is notable because the federal court rejected the general applicability of the balancing test used in the *Mountain Copper* decision:

> The rule that the comparative convenience or inconvenience of the parties from granting or withholding an injunction when it is sought to abate a nuisance . . . must . . . be confined to the particular case then before the court [*Mountain Copper*], and is not to be accepted as the statement of a rule to be applied generally in cases where it is sought to abate a nuisance, the existence of which is undoubted. (*American Smelting and Refining Co. v. Godfrey* 1907, 230)

There were a number of responses to the decision, including complete shutdown, relocation of operations, and installation of abatement technologies. Bingham Consolidated and Utah Consolidated abandoned operations and moved to Tooele, Utah, in 1908. American Smelting and Refining Company (ASARCO), rebuilt its flue and dust chamber, and also added a bag house at its lead smelter. U.S. Smelting also added a bag house at its Midvale operation, and the company did not reopen its copper smelter (Fulton 1915, 82–83).

The continuation of operations was coordinated through an agreement with the smelters and the farmers' association. The smelters agreed to not treat ore containing more than 10 percent sulfur and to prevent arsenic releases. Compliance with these agreements required state-of-the-art technology. Although bag houses at lead smelters had become economically viable by this time, the

Utah lead operations installed them only after being forced to do so by the court. At this time, installation of bag houses at copper smelters was not the industry norm; in fact, there was only one operating bag house at a copper smelter by 1914 (Fulton 1915, 57).

Plaintiffs could ask the court to enforce the injunction if terms of agreement were not satisfied. The smelters found themselves back in court as the result of a 1916 suit that enjoined the Murray and Midvale smelters (*Anderson v. American Smelting and Refining Co.* 1919). The injunction mandated that the smelters reduce their sulfur dioxide emissions, and the plaintiffs were awarded damages.

In California, a state court enjoined the Selby Smelting and Lead Company in 1908, prohibiting smelting operations during the growing season (*California v. Selby Smelting & Lead Co.* 1908). The company subsequently installed a bag house and an ESP, and began shipping its high-sulfur ores elsewhere (Fulton 1915, 86–87). But area farmers continued to complain, leading to the formation of the Selby Smelter Commission (Holmes, Franklin, and Gould 1915).[19]

Also in California, the Shasta County Farmers' Protective Association asked for injunctive relief against Balaklala Consolidated in Coram and against Mammoth Copper in Kennett. In 1910 the smelters and the farmers came to terms for continued smelter operations. In response, Mammoth Copper became the first copper smelter to put in a bag house. The Balaklala Consolidated smelter also installed fans and an ESP, but failure to capture sulfur dioxide eventually led to the shutdown of the smelter. Again, injunctive relief forced the firm to adopt abatement technology.

Denial of Relief

In Montana, the Anaconda company located its Washoe smelter in the Deer Lodge Valley because of easier access to timber and water resources than what was available in Butte. Anaconda paid $330,000 to farmers in surrounding areas in 1902, and in exchange the farmers waived all future claims. Anaconda's offer to purchase these "smoke easements" looks like the first step of a bargaining process that would serve as an alternative to litigation, but there appears to have been little negotiation between the farmers and Anaconda beyond the initial offers. Instead, Anaconda advised farmers who were not willing to accept the offer that their options were to "take it [the initial offer] or go to court" (MacMillan 1973, 113–14).

The plaintiffs were certainly aware that litigation would be a costly enterprise (though they perhaps were not aware that court damage awards were generally much less than actual damages). Even so, the ensuing litigation indicates that the compensation offered by Anaconda, indeed, was insufficient. In 1905 the Deer Lodge Valley Farmers' Association demanded $1,175,000 to cover damages to land, water rights, and improvements of 60,525 acres, more than 90 percent of the agricultural land in the valley. When the company refused to pay, Fred Bliss, a citizen of Idaho, filed for an injunction in federal court on behalf of the farmers' association. At the time Washoe was responsible for more than 10 percent of world smelter output. In 1909 the court denied the injunction and awarded $350 to Bliss.[20] The federal appeals court also denied the injunction.

President Theodore Roosevelt and the Department of Justice had been watching the Anaconda case since it began and had tried to mediate a settlement as early as 1908 without success. Late in his term, Roosevelt directed the Department of Justice to begin preparing for a possible federal suit against the Anaconda company on the basis of damage to the surrounding national forests. With only a short time left in his administration, Roosevelt decided not to proceed with the suit but instead to leave it in the hands of his successor, William Howard Taft. By threatening Anaconda with a federal suit, Taft was able to get the company to agree to abide by the recommendations of the Anaconda Smelter Commission, which began studying the issue in 1911.

The company eventually installed a small-scale ESP to capture gold, silver, and copper in 1919, and put in a large-scale ESP to capture arsenic in 1923. The company did not build a large-scale sulfur dioxide plant until after World War II. It could have set up an acid plant to reduce sulfur emissions by 1910, but claimed it would not have been a profitable enterprise and balked at the implementation. Thus, the settlement slowed the adoption of the pollution control measures.

CONCLUSION

This chapter has examined the incentives for firms to adopt pollution-abatement technologies. The focus has been on nuisance doctrine and the court's choice of injunctive relief versus an award of compensatory damages. Of interest is the historical move to a balancing test that attempted to assess the benefits and costs of abating pollution. The introduction of balancing tests made it more likely that compensatory damages would be awarded in any given nuisance suit.

Although many scholars argue that the balancing test is an efficient element of nuisance doctrine, there are a number of reasons to question this premise in the context of U.S. mining. Compensatory damages typically far understated the true level of damages, meaning that court remedies did not provide the appropriate incentives for industrial polluters to reduce the off-site impacts of their operations. Moreover, when compensatory damages were awarded instead of an injunction, the effect was to retard the adoption of pollution-abatement technologies. Low damage awards, the low probability of successfully petitioning for an injunction, and high court costs all dampened the incentive for potential plaintiffs to sue polluters.

When an injunction was issued, there were often observable reductions in pollution. In some cases, smelting operations shut down altogether. If the smelter reopened, it was bound to comply with stipulations of the injunction to mitigate its offsite impacts. This allowed for long-term agreements between the plaintiff and the defendant even when a large number of parties (e.g., dozens of farmers and several smelters) were involved.

The courts are a costly and time-consuming way to establish property rights, and the damages that accrued during the litigation process were often substantial. Yet in almost every case, we find the pioneering advance in pollution-abatement technologies at smelters where a court order clarified and protected property rights by mandating a reduction in emissions. A review of copper smelting during the early part of this century suggests that courts encouraged the development and installation of new pollution-abatement technologies.

Although Posner (1992) suggests that the enactment of environmental statutes has displaced nuisance for addressing pollution problems, a more plausible story is that the intrusion of a balancing test into nuisance law may well have made statutory measures necessary to protect potential plaintiffs from environmental harms. Recent research by the Environmental Law Institute (1998) shows that barriers to innovation in pollution-abatement technologies are the result of the design and enforcement of environmental statutes and regulations. Thus, the role of the courts and the choice of legal remedies will continue to be a fruitful area of research.

NOTES

1. Calabresi and Melamed (1972). There is now a substantial literature concerning the efficient and equitable choice of a legal remedy. Economic analysis of the rules is found in Polinsky (1980a, 1980b) and Kaplow and Shavell (1996). The 1997 *Yale Law Journal* contained a number of articles commemorating the twenty-fifth anniversary of the Calabresi and Melamed piece, including Epstein (1997).

2. The important feedback effects between the institutional and industrial structure are explored in David and Wright (1997). For an excellent discussion of the interrelationships between technology and economic progress, see Mokyr (1990).

3. A public choice explanation is that balancing tests increase the discretion of the court, and the expectation is that judges will craft decisions that enhance the power of the court.

4. Both Horwitz and Posner, however, have a pessimistic view of the courts' effectiveness of dealing with environmental problems. See Meiners and Morriss (2000) for the alternative view.

5. Our discussion of the influence of legal rules assumes that the development of abatement technology was a social good, that is, the benefits of developing and using the technology outweighed the cost.

6. Although American mining interests were aware of several rich copper deposits in the West, high transportation costs retarded developments until the early 1880s. An exception was the northern Michigan copper region, which enjoyed the benefits of Great Lakes shipping. During the 1860s and 1870s, Calumet and Hecla used capital provided by investors centered primarily in Boston to generate the first U.S. copper boom (Navin 1978, 12). Nevertheless, the first major western copper operations began production in 1881 without ready access to railroads. At Butte and Bisbee, located in the future states of Montana and Arizona, respectively, miners began with surface deposits of silver and gold. As the mines sank deeper, however, the Butte and Bisbee miners discovered increasingly large deposits of copper. Initial gold and silver production at these sites provided financing for the capital needed to process and transport copper (Navin 1978, 13).

7. Although prices were quite volatile over this period, the difference between the wholesale price index in 1881 and in 1910 was negligible.

8. Another early adoption of an acid plant was at Mountain Copper in California. In that case, however, the company chose to abort the project because acid production was not a profitable enterprise. This is discussed in the following section.

9. There is a distinction between a public and a private nuisance. A public nuisance affects the health, safety, and the community in a fairly uniform fashion. Private citizens have no remedy for a public nuisance unless an individual suffers some harm that is substantially different from how it affects the "average member" of the community. The distinction had substantial impacts on whether citizens could redress their grievances through the courts: "The most significant fact about the public nuisance doctrine was that it enabled courts to extend to private companies virtually the same immunity from lawsuits that the state received under the theory of consequential damages" (Horwitz 1977, 76–77; see also Brenner 1974, 420–24). In fact, some state legislators tried to protect copper interests by declaring certain smelter operations to be public uses. See, for instance, the court's discussion of the Utah statute in *American Smelting and Refining Co. v. Godfrey* (1907).

10. The Coase (1960) theorem demonstrates that when bargaining costs are low and the assignment of rights is clear, the property right will flow to the party that places the highest value on the right. As a result, the wealth-maximizing outcome occurs independently of who initially owns the right (in the cases we are dealing with, such wealth includes long-term prospects for soil, water, and air quality, however valued). As a result, the assignment of the right should have no bearing on the outcome, though it certainly affects the distribution of wealth. Although law and economics scholars believe that this is a reasonable proposition, Farnsworth (1999) examined twenty cases and found no instances of bargaining after the injunction was issued.

11. Of course, holdout situations also occur when there are a large number of bargaining parties, when one or more parties acts strategically in an attempt to capture a disproportionate share of the settlement.

12. See Epstein (1997) for a strong critique of this argument.

13. Kaplow and Shavell (1996) argue that compensatory damages are efficient when there are many damaged parties, even if the court is unlikely to accurately assess damages. The authors, however, only briefly discuss how property rules and liability rules affect the development of pollution-abatement technologies (725). Moreover, they concede that their conclusion regarding the superiority of liability rules for dealing with pollution externalities does not hold if courts systematically underestimate damages (730–32).

14. Brenner (1974) and Nedelsky (1981) describe similar legal climates in England and Canada, respectively.

15. Brenner (1974) cites three cases between 1857 and 1865 in which courts granted injunctive relief in cases involving air and water pollution.

16. For instance, in *Madison v. Ducktown Smelter* (1904) in Tennessee, the court awarded the plaintiff approximately $1,000 after the smelter operation rendered his land virtually worthless. Dewees (1995) estimates property damages at $255,000 for land that was decimated, plus an additional $42,000 for other land that was affected but not rendered worthless. In addition, he estimates health costs in the area associated with arsenic pollution at more than $1 million. Dewees finds similar imbalances between actual damages and court awards for cases across the United States and Canada.

17. The discussion includes all the cases in which new technologies were adopted in response to litigation. We do not discuss a number of famous cases, including the *Trail Smelter* that led to an international dispute between the United States and Canada.

18. It is still possible that denial of the injunction was efficient in that (1) the harm to the defendant exceeded harm to the plaintiff; and (2) granting the injunction would not have resulted in the movement of property rights to the high-value user.

19. In several instances, the result of litigation was the formation of a special commission to investigate means of addressing the problem.

20. *Bliss v. Anaconda Copper Mining Co. et al.* (1909). Dewees (1995) estimates property damage at $2,417,885 and health damages at more than $1 million.

REFERENCES

Barger, Harold, and Sam H. Schurr. 1944. *The Mining Industries, 1899–1939: A Study of Output, Employment, and Productivity.* New York: American Book-Stratford Press.

Brenner, Joel F. 1974. Nuisance Law and the Industrial Revolution. *Journal of Legal Studies* 3: 403–33.

Calabresi, Guido, and A. Douglas Melamed. 1972. Property Rules, Liability Rules, and Inalienability: One View of the Cathedral. *Harvard Law Review* 85(6): 1089–128.

Cameron, Frank. 1952. *Cottrell: Samaritan of Science.* New York: Doubleday & Company.

Coase, Ronald. 1960. The Problem of Social Cost. *Journal of Law & Economics* 3(1): 1–44.

David, Paul A., and Gavin Wright. 1997. Increasing Returns and the Genesis of American Resource Abundance. *Industrial and Corporate Change* 6(2): 203–45.

Dewees, Donald N. 1995. Sulfur Dioxide Emissions from Smelters: The Historical Inefficiency of Tort Law. Working paper, RePEc:tor:tolaec:S44. University of Toronto, Law and Economics Program.
Encyclopedia of Chemical Technology. 1977. New York: John Wiley & Sons.
Environmental Law Institute. 1998. *Barriers to Environmental Technology Innovation and Use.* Environmental Law Institute Research Report. Washington, D.C.
Epstein, Richard. 1997. A Clear View of the Cathedral: The Dominance of Property Rules. *Yale Law Journal* 106: 2091–120.
Farnsworth, Ward. 1999. Do Parties to Nuisance Cases Bargain after Judgment? A Glimpse Inside the Cathedral. *University of Chicago Law Review* 66: 373–436.
Fells, James E. 1979. *Ores to Metals: The Rocky Mountain Smelting Industry.* Lincoln: University of Nebraska Press.
Franzle, Otto. 1993. *Contaminants in Terrestrial Environments.* Berlin: Springer-Verlag.
Fulton, Charles H. 1915. *Metallurgical Smoke.* U.S. Bureau of Mines Bulletin 84. Washington, D.C.: Government Printing Office.
Hellenbeck, William H., and Kathleen M. Cunningham-Burns. 1985. *Pesticides and Human Health.* New York: Springer-Verlag.
Holliday, J. S. 1999. *Rush for Riches: Gold Fever and the Making of California.* Berkeley: Oakland Museum of California and University of California Press.
Holmes, J. A., Edward C. Franklin, and Ralph A. Gould. 1915. *Report of the Selby Smelter Commission.* U.S. Bureau of Mines Bulletin 98. Washington, D.C.: Government Printing Office.
Horwitz, Morton J. 1977. *The Transformation of American Law, 1780–1860.* Cambridge, Mass.: Harvard University Press.
Kaplow, Louis, and Steven Shavell. 1996. Property Rules versus Liability Rules: An Economic Analysis. *Harvard Law Review* 109: 713–90.
Kelley, Robert L. 1959. *Gold versus Grain: The Hydraulic Mining Controversy in California's Sacramento Valley. A Chapter in the Decline of the Concept of Laissez Faire.* Glendale, Calif.: Arthur H. Clark.
LeCain, Timothy J. 1998. Moving Mountains: Technology and the Environment in Western Copper Mining. Ph.D. dissertation, University of Delaware, Newark.
MacMillan, Donald. 1973. A History of the Struggle to Abate Air Pollution from Copper Smelters in the Far West, 1885–1933. Ph.D. dissertation, University of Montana, Missoula.
Meiners, Roger E., and Andrew P. Morriss, eds. 2000. *The Common Law and the Environment: Rethinking the Statutory Basis for Modern Environmental Law.* Lanham, Md.: Rowman & Littlefield, Publishers.
Mokyr, Joel. 1990. *The Lever of Riches: Technological Creativity and Economic Progress.* New York: Oxford University Press.
Navin, Thomas R. 1978. *Copper Mining and Management.* Tucson: University of Arizona Press.
Nedelsky, Jennifer. 1981. Judicial Conservatism in an Age of Innovation: Comparative Essays on the Canadian Nuisance Law. In *Essays in the History of Canadian Law,* vol. 1, ed. David H. Flaherty. Toronto: Osgood Society, 281–322.
Peters, Edward D. 1911. *The Practice of Copper Smelting.* New York: McGraw-Hill.
Polinsky, A. Mitchell. 1980a. On the Choice between Property Rules and Liability Rules. *Economic Inquiry* 18: 233–46.

———. 1980b. Resolving Nuisance Disputes: The Simple Economics of Injunctive and Damage Remedies. *Stanford Law Review* 32: 1075–112.

Posner, Richard A. 1992. *Economic Analysis of Law*, 4th ed. Boston: Little, Brown & Company.

Quinn, M. L. 1993. Industry and Environment in the Appalachian Copper Basin, 1890–1930. *Technology and Culture* 34(3): 575–612.

Rosen, Christine. 1993. Differing Perceptions of the Value of Pollution Abatement across Time and Place: Balancing Doctrine in Pollution Nuisance Law, 1840–1906. *Law and History Review* 11(2): 303–82.

Swain, Robert. 1949. Smoke and Fume Investigations. *Industrial and Engineering Chemistry* 41(11): 2384–87.

Swenson, Robert W. 1968. Legal Aspects of Mineral Resources Exploitation. In *History of Public Land Law Development*, ed. Paul W. Gates, ch. 23. Washington, D.C.: Zenger.

Tver, David F. 1981. *Dictionary of Dangerous Pollutants.* New York: Industrial Press.

CASES CITED

American Smelting and Refining Co. v. Godfrey, 158 F. 225 8th Circuit (1907).

Anderson v. American Smelting and Refining Co., 265 F. 928 (D. Utah 1919).

ASARCO v. Godfrey, 158 F. 225 (1907).

Bliss v. Anaconda Copper Mining Co. et al., 167 F. 342 (D. Montana 1909); aff'd 186 F. 789 (9th Cir. 1911).

California v. Selby Smelting & Lead Co., Sup.Ct. Solano Cty., July 16, 1908 (unreported).

Georgia v. Tennessee Copper Co. and Ducktown Sulphur, Copper and Iron Company, Limited, 237 US 474 S.Ct. 631 (1915).

Madison v. Ducktown Sulphur, Copper and Iron Co., 113 Tenn. 331, 83 S.W. 658 (1904).

Mountain Copper v. United States, 142 F 625 9th Cir. (1906).

Woodruff v. North Bloomfield Gravel and Mining Co., 18 Fed. 753, 768, 806 (1884).

Index

About the Political Economy Forum and the Authors

PERC is a nonprofit research center located in Bozeman, Montana, that focuses on market solutions to environmental problems. For two decades PERC has been a pioneer in recognizing the importance of markets, personal initiatives, and property rights in promoting environmental quality. This approach is known as free market environmentalism. PERC associates have applied this approach to a variety of issues, including resource development, water marketing, chemical risk, private provision of environmental amenities, public land management, and endangered species protection.

In 1989 PERC organized the first of an ongoing series called the Political Economy Forum aimed at applying the principles of political economy to important policy issues. Each forum brings together scholars in economics, political science, law, history, and other disciplines to discuss and refine academic papers that explore new applications of political economy to policy analysis.

The chapters in this volume emanate from the Political Economy Forum, "The Technology of Property Rights," held in December 1999. The forum was organized and directed by Terry L. Anderson and Peter J. Hill. In 1975 they published an article titled "The Evolution of Property Rights: A Study of the American West" in which they developed a model for explaining how property rights evolve. One component of that evolution, technology, was illustrated by the impact of the invention of barbed wire on the development of property rights to land and livestock. This forum goes beyond historical examples and considers how technology can affect the evolution of property rights to fish, wildlife, water, and pollution. To the extent that technology can lower the costs of defining and enforcing property rights, market solutions to environmental problems become more feasible and political solutions less necessary.

POLITICAL ECONOMY FORUM PARTICIPANTS

The participants were as follows: Jonathan Adler, Competitive Enterprise Institute; Bruce Benson, Florida State University; Richard Cahoon, Cornell Research Foundation; William Dennis, Liberty Fund, Inc.; David Haddock, Northwestern University; Daniel Klein, Santa Clara University; Gary Libecap, University of Arizona; Dean Lueck, Montana State University; Roger E. Meiners, PERC and University of Texas, Arlington; Andrew P. Morriss, Case Western Reserve University; Kathryn Ratté, Foundation for Teaching Economics; Richard L. Stroup, PERC and Montana State University; and Ian Wills, Monash University.

AUTHORS

Terry L. Anderson is executive director of PERC in Bozeman, Montana, senior fellow at the Hoover Institution, Stanford University, and professor of economics (retired) at Montana State University. His work has helped launch the idea of free market environmentalism and has prompted public debate over the balance between markets and government in managing natural resources. Anderson has been published widely in professional journals as well as the popular press and is author or editor of twenty-one books, including *Political Environmentalism* and *Enviro-Capitalists: Doing Good While Doing Well*, the latter coauthored with Donald Leal. Anderson received his Ph.D. in economics from the University of Washington and has been a visiting scholar at Oxford University, the University of Basel, and Cornell University Law School. He was awarded a Fulbright research fellowship to Canterbury University and was a national fellow at the Hoover Institution.

Gregory B. Christainsen is professor of economics at California State University at Hayward. He received his Ph.D. in economics from the University of Wisconsin-Madison. Christainsen has authored or coauthored more than fifty articles in scholarly books and journals, including the *American Economic Review*, *Journal of Environmental Economics and Management*, *Natural Resources Journal*, *The Environmental Professional*, *The ANNALS of the American Academy of Political and Social Science*, *Quarterly Journal of Economics and Business*, *Public Finance Quarterly*, *Cato Journal*, *Journal of Private Enterprise*, *Journal of Economic Education*, and the North-

Holland Publishing Company's *Handbook in Natural Resource and Energy Economics*. His research affiliations have included the University of Wisconsin-Madison, Harvard Law School, the Organization for Economic Cooperation and Development (OECD, Paris), the Office of Technology Assessment of the U.S. Congress, and the Reserve Bank of New Zealand. He was formerly the research director and senior economist for the Pacific Research Institute in San Francisco. He has lectured or worked as a consultant in more than thirty countries.

David Gerard is a postdoctoral fellow in the department of engineering and public policy at Carnegie Mellon University and a former research associate at PERC. He earned a Ph.D. in economics from the University of Illinois at Urbana-Champaign, with specialization in natural resource and environmental economics, law and economics, and American economic history. Gerard has published articles addressing public lands, the causes and consequences of federal mineral land policies, the development of the federal wilderness system, and the use of market mechanisms to encourage compliance with environmental statutes.

Brian C. Gothberg is completing his master's in economics at California State University at Hayward. His areas of research include natural resources and urban transportation. He is also the principal economic analyst at the University of California's Industry-University Cooperative Research Program in Berkeley. Gothberg has been a computer programmer and edited the second volume of the University of California's acclaimed "Family Math" series, *Family Math for Young Children* (1997).

Peter J. Hill is professor of economics at Wheaton College, where he holds the George F. Bennett chair. He is a senior associate of PERC, and in addition to this book, has coedited six previous volumes of the Political Economy Forum series. His research and articles, especially on the evolution of property rights in the American West, helped found the new resource economics that is the basis for free market environmentalism. He is coauthor, with Terry L. Anderson, of *The Birth of a Transfer Society* and, with Joseph L. Bast and Richard C. Rue, of *Eco-Sanity: A Common Sense Guide to Environmentalism*. As an economic consultant he has worked with the Bulgarian government in its attempts to privatize agricultural lands. Hill has a B.S. from Montana State University and a Ph.D. from the University of Chicago.

Daniel Huppert is associate professor of marine affairs at the University of Washington, Seattle, where his teaching includes economics of marine policy and marine resources management. After completing Ph.D. work at the University of Washington, he worked as an economist and program leader for the National Marine Fisheries Service in La Jolla, California, from 1974 to 1989. While working at the fisheries service, he gave advice on U.S. tuna policy, led the development of fishery management plans for coastal pelagic species in California (anchovies, mackerel, squid), participated on committees for the Pacific Fishery Management Council, and conducted economics research of marine recreational fishing. He chaired the Scientific and Statistical Committee for the Fisheries Council on two occasions, authored numerous technical reports, and published papers concerning economics and fisheries management. He served on the North Pacific Fishery Management Council's Scientific and Statistical Committee (1990–94), chaired NMFS's Economics Technical Committee on Snake River salmon recovery (1992–96), and is currently chair of the Northwest Power Planning Council's Independent Economic Analysis Board. Research interests include social aspects of Pacific Northwest coastal ecosystems management, salmon conservation and recovery, economics of climate change and climate forecast information, and economics of recreation.

Gunnar Knapp is a professor of economics at the University of Alaska Anchorage Institute of Social and Economic Research, where he has worked since receiving his Ph.D. in economics from Yale in 1981. He has conducted research on a wide variety of Alaska resource management issues and is a specialist on Alaska fisheries management and fisheries markets. Knapp has conducted surveys of Alaska fishermen on effects of the introduction of individual fishing quota management in the Alaska halibut fishery, and attitudes of fishermen toward this management.

Clay J. Landry is a research associate at PERC in Bozeman, Montana, and research director for Optimal Water, Inc., a water brokerage firm based in Dana Point, California. He is the author of *Saving Our Streams through Water Markets: A Practical Guide.* Landry has been published in several professional journals and the popular press, including the *Wall Street Journal, Water Resources Impact, Orange County Register,* and numerous western regional newspapers. He is also a regular contributor to *Global Water Intelligence,* an international water industry newsmagazine. Landry holds a master's degree in agriculture and resource economics from Oregon State University and a bachelor's degree in economics from the University of Wyoming. He first became interested in water policy in his youth while helping to irrigate his parents' small Montana farm.

Timothy J. LeCain is a historical consultant and an adjunct professor of history at Montana State University. He received a Ph.D. in the history of technology from the University of Delaware in 1989 and has been a fellow at the Hagley Museum, the Smithsonian Institution, and the Huntington Library. His recently published articles have analyzed the historical limits of eco-efficiency in western copper smelting and the technology of large-scale open-pit mining. LeCain is writing a book that examines the environmental effects of mass production techniques in mining and other extractive industries.

Anna M. Michalak is a Ph.D. candidate in the Department of Civil and Environmental Engineering at Stanford University. Her research focuses on developing stochastic inverse modeling methods for contaminant source identification, and determining the effectiveness of such tools given the scientific, economic, and political settings of specific case studies. She completed her undergraduate studies at the University of Guelph, Ontario, Canada, where she received a B.Sc.(Eng.) in environmental engineering. After moving to California she earned an M.S. degree in environmental fluid mechanics and hydrology from the Department of Civil and Environmental Engineering at Stanford University. Michalak has been named a Stanford graduate fellow and Gabilan fellow by Stanford University.

Robert B. Naeser is a water resource specialist with Science Applications International Corporation (SAIC) in Denver, Colorado. He works on a variety of water resource projects for local, state, and federal government clients. Typical water resource projects he has been involved with include water rights analysis, Clean Water Act enforcement cases, natural resource damage assessments, water supply and wastewater analysis, environmental permitting, and water quality analysis for mining and transportation projects. Naeser served as a consultant to the National Audubon Society in 1996 and the Sierra Club Legal Defense Fund in 1993 for water-related development issues. He holds a B.A. in economics (1992) from The Colorado College and an M.E.S. from Yale University (1997), where he studied water resource economics and management. Naeser is an alumnus of PERC's summer symposium for undergraduates on free market environmentalism.

Mark Griffin Smith is an associate professor of economics at The Colorado College, where he is chair of the economics and business department. His research interests include water markets, instream flow protection, reliability in water supply systems, and water quality management in Central and Eastern Europe. He has worked for Appropriate Technology, the President's Nuclear Safety Oversight Committee, and the World Bank and taught at Duke University, where he received his Ph.D. From 1994 to 1996 he was project manager and research scholar with the Water Resources Project at the International Institute for Applied Systems Analysis (IIASA) in Laxenburg, Austria. In 1999 he was appointed the first John L. Knight chair for the Study of Free Enterprise.

Barrett P. Walker serves as a trustee for the Alex C. Walker Educational and Charitable Foundation. The purpose of the foundation is to address issues related to the causes of economic imbalances. In this position he has sought to further free market approaches to conserving environmental resources. Prior to becoming a foundation officer, he pursued consecutive careers as an instructor in ophthalmology at the Baylor College of Medicine and Emory University and a small-business owner. Walker's interest in economics and the environment led him to leave the world of small business in the early 1990s to pursue a master's degree in ecology and sustainable development from the University of Georgia Institute of Ecology. Subsequent to his graduation, Walker chaired the Metropolitan Atlanta Watershed Initiative in 1998.

Bruce Yandle, a PERC senior associate, is alumni distinguished professor of economics and BB&T scholar at Clemson University. Yandle received an A.B. from Mercer University and a Ph.D. from Georgia State University. He served as senior economist on the President's Council on Wage and Price Stability and as executive director of the Federal Trade Commission. Yandle is author or editor of many books including *The Political Limits of Environmental Regulation*, *Common Sense and Common Law for the Environment*, and *Land Rights: The 1990s' Property Rights Rebellion.*

www.ingramcontent.com/pod-product-compliance
Ingram Content Group UK Ltd.
Pitfield, Milton Keynes, MK11 3LW, UK
UKHW021034150926
475561UK00003B/104